Consuming Interests

The position of food in advanced societies is currently being redefined. As governments are struggling to deal with both the emergent food scares and new technological advances, consumers become more sensitive to the quality and provenance of foods. At the same time, corporate retailers have significantly increased their grip upon the co-ordination and provision of food supply, quality and choice.

Consuming Interests focuses on the key interests which lead to the provision of food choices: corporate retailers, government regulators and consumer organisations; and examines how a retail-led form of food governance has emerged. Divided into three sections: Concepts and Framework, National Strategies and Local Strategies, the book provides a detailed examination of corporate retailers, state agencies and consumer organisations involved in the food sector. The analysis raises some key social scientific questions concerning what the public can expect the central and local state to ensure, what limits there may be upon state action, and what the most appropriate balances should be between public and private interests in the provision of 'quality' foods.

Blending critical theory, empirical research and policy, *Consuming Interests* provides a topical and interdisciplinary exploration into the nature of food provision, policy and regulation and gives an insight into the broader social science concerns of the nature and powers of the contemporary state.

Terry Marsden is Professor of Environmental Policy and Planning at Cardiff University. **Andrew Flynn** is Senior Lecturer in Environmental Policy and Planning at Cardiff University, **Michelle Harrison** is a researcher and consultant at the Henley Centre in London.

Consumption and space

Series editors:
Peter Jackson
University of Sheffield
Michelle Lowe
University of Southampton
Frank Mort
University of Plymouth

Adopting an inter-disciplinary perspective, and combining contemporary and historical analysis, *Consumption and space* aims to develop a dialogue between cultural studies and human geography, opening up areas for serious intellectual debate.

Published

Hard looks
Masculinities, spectatorship and contemporary consumption
Sean Nixon

Material cultures: why some things matter
Daniel Miller (editor)

Consuming interests
The social provision of foods

Terry Marsden, Andrew Flynn
and Michelle Harrison

First published 2000 in the UK and the USA
by UCL Press
11 New Fetter Lane, London EC4P 4EE

The name of University College London (UCL) is a registered trade
mark used by UCL Press with the consent of the owner.

UCL Press is an imprint of the Taylor & Francis Group

Typeset in Garamond by Routledge
Printed and bound in Great Britain by Biddles Ltd,
Guildford and King's Lynn

British Library Cataloguing in Publication Data
A catalogue record for this book is available from the British Library

Library of Congress Cataloging in Publication Data
Marsden, Terry
Consuming interests: the social provision of foods/Terry Marsden
Andrew Flynn, Michelle Harrison
p. cm. – (Consumption and space)
Includes bibliographical references and index.
1. Food industry and trade–Social aspects. 2. Food contamination
and inspection. 3. Food law and legislation. 4. Food–marketing.
5. Consumer protection. I. Flynn, Andrew. II. Harrison, Michelle.
III. Title. IV. Series.
HD9000.5.M3581999
338.4'7664–dc219927329
CIP

ISBN 1–857–28899–8 (hbk)
ISBN 1–857–28900–5 (pbk)

Contents

Illustrations

Figures

Tables

Preface

Food and the state, the state of food

The position of food in advanced societies is currently being redefined. This is occurring at a time when broader social and political changes suggest a growing significance of consumption as an active process of material life, and when modern states become increasingly conscious of the need to provide opportunities for choice. This book focuses upon the key (consuming) interests which lead to the provision of food choices – corporate retailers, government regulators and consumers' organisations – and it examines how, in the UK at least, a retail-led form of food governance has emerged.

As governments continue to struggle to deal with both the emergent food scares and the new technological advances being made (most recently seen in the debates surrounding bovine spongiform encephalopathy (BSE), the proposal for a new Food Standards Agency and the arrival of genetically modified organisms (GMOs)), and consumers become, partly as a consequence, much more sensitive to the quality and provenance of foods, corporate retailers have significantly increased their grip upon the coordination and provision of food supply, quality and choice. This raises some central social scientific questions concerning what the public can expect the central and local state to provide and ensure, what limits there may be upon state action, and what the most appropriate balances should be between public and private interests. How can the state represent the public interest in such a privatised and technically sophisticated world? Are there new models of state–private sector governance emerging which can better deliver public benefits? How is the arena of food consumption to be regulated in ways which appease both consumer and corporate interests? What is the nature of retail competition in providing choices to consumers?

These broad questions provide the canvas for this work, and they particularly highlight the two related themes of this series: consumption and space. Space and spatiality are central, in our view, to the analysis and understanding of these broad social and political questions, not simply as outcomes but also as complex ingredients in the provision of foods. We show how, for instance, retail restructuring and government regulation and policy have to constantly interact at strategic (mainly EU and national) and local levels.

While neither are completely set in stone, both have to be incorporated in corporate strategy and policy development and implementation. Retailers have to prise open 'competitive spaces', while regulators face the almost imposs- ible task of ensuring 'uniform' implementation of policy across different 'consumption spaces'. The provision and consumption of foods is inherently spatial, and this spatiality has to be defined and dealt with constantly by both public and private sector interests in a world of increasingly 'careful consumption'.

This book explores these questions. It represents the continued debates two of the authors (Marsden and Flynn) initiated from the early 1990s, concerning the character of the British food sector and the significance of the food scares (particularly BSE) in shaping the current regulation of food supply. These early debates were assisted by postgraduates and researchers, Ian Drummond, Neil Ward, Sarah Williams and Marion Justice. This led to a major Economic and Social Research Council (ESRC) project, entitled *Retailing, regulation and food quality: the social construction of the consumer interest*, Nations' Diet Programme (1994–7). It is the research funded under this study which is reported in this volume, and we wish to acknowledge not only the financial support of the ESRC, but also the central role that publicly funded independent research organisations can play in protecting and maintaining social science research in those areas (such as food policy and retailing) of both considerable public and commercial sensitivity.

A large number of private and public representatives gave the research team (Marsden, Flynn and Harrison) a considerable amount of valuable access, including retailers, government officials and members of food organisations (both in the UK and EU). The research could not have been completed without their assistance. In particular we are grateful to the local enforce- ment officials in our study areas who spent large amounts of time in helping us to complete the ethnographic parts of the work. In addition, we are grateful to Sally Harris, who helped us to administer much of the early part of the work, and to Joek Roex, who has maintained the research office in good shape while also catering for the considerable and idiosyncratic demands of the three authors. We would also like to acknowledge the always thoughtful, encouraging and critical advice continually given in our Nations' Diet Programme meetings with its director, Anne Murcott. We would also wish to acknowledge the continuous support in all sorts of ways of Mary Anne, Joseph and Hannah Marsden and Suki, Jake, Laura and Holly Flynn.

Dealing centrally with the interactions between food retailers, regulation and consumption, the attempt here is to integrate some of the more signifi- cant concepts and theories from the disparate disciplines of retail geography, political science, and the sociology and geography of consumption in ways which allow a more comprehensive understanding of the nature of food provision. In doing so, we have, in compiling this text, also been aware of how such a focus upon the food sector, and particularly upon some of its key actors, can reflect broader changes in the nature of state–society relationships. The very obsolescence of either of the postwar dominant philosophies of traditional hyper-welfarist or neo-conservative deregulation is clearly visible in the contemporary study of food. New forms of regulation – a process of re-regulation – becomes a defining characteristic and outcome of the current shape of consuming interests.

Terry Marsden
Andrew Flynn
Michelle Harrison
Cardiff, November 1998

Abbreviations

BEUC	Bureau Européen des unions des consommateurs (European Consumers Organization)
BIGA	British Independent Grocers Association
BNF	British Nutrition Foundation
BRC	British Retail Consortium
BSE	bovine spongiform encephalopathy
BST	bovine somatotropin
CA	Consumers Association
CIAA	Confédération des industries agro-alimentaires de l'UE (Confederation of the Food and Drink Industries of the EU)
CCP	critical control point
CEG	Consumers in Europe Group
CJD	Creutzfeldt Jacob disease
COMA	Committee on the Medical Aspects of Food Policy
DG III	EU Directorate General III (Industry)
DG VI	EU Directorate General VI (Agriculture)
DG XXIV	EU Directorate General XXIV (Consumer Policy and Consumer Health Protection)
DoE	Department of the Environment
DoH	Department of Health
DTI	Department of Trade and Industry
ECR	Court of Justice of the European Communities, *Reports*
EEC	European Economic Community
EFTA	European Free Trade Area
EHO	Environmental Health Officer
EU	European Union
FC	Food Commission
FDF	Food and Drink Federation
GLC	Greater London Council
GM	genetically modified
GMO	genetically modified organism
HACCP	hazard analysis and critical control point

IGD	Institute of Grocery Distribution
ISO	International Organization for Standardization
LACOTS	Local Authorities Coordinating Body on Food and Trading Standards
MAFF	Ministry of Agriculture, Fisheries and Food
MLC	Meat and Livestock Commission
MP	Member of Parliament
NCC	National Consumer Council
NFA	National Food Alliance
NFCG	National Federation of Consumer Groups
NFU	National Farmers' Union
NHS	National Health Service
OFT	Office of Fair Trading
OJ	*Official Journal of the European Communities*
PPG	Planning Policy Guidance
SCF	European Commission Standing Committee on Foodstuffs
SEAC	Spongiform Encephalopathy Advisory Committee
TSO	Trading Standards Officer

1 Introduction

Regulation, retailing and consumption: deregulating states and concerned consumers

Longstanding questions about how and why states regulate and what they should regulate have been given renewed concern by governments in Britain and America since the early 1980s. Previously well-accepted public policy goals and delivery mechanisms have been opened up to serious scrutiny, and in a number of cases subjected to reform. It is important, though, to distinguish between the rhetoric of politicians and those who clamour round them, and the actions of this period. The implications of the policies of the 'deregulating' governments at this time are by no means clear-cut; indeed, it is only with the progress of time that we can fully begin to assess the changes that have taken place to the nature of the state, and to disentangle the impacts of particular governments from longer term processes of restructuring.

Of course, processes of state restructuring are very broad, and our particular interest here relates to regulation and more specifically the regulation of food in Britain from the early to late 1990s. This period of time proved to be particularly propitious in that not only was it the culmination of an extended period of Conservative government, thus allowing us to evaluate the nature and extent of regulatory change, but it also marked a time of considerable change in food retailing. The 1990s have witnessed an unprecedented intensity of corporate retailer competition, simultaneous with the maintenance of a small but differentiated group of retailers dominating the British market. The latter account for over 44 per cent of grocery sales and 67 per cent of packaged groceries. At the same time, however, national consumer and government concerns about the quality of foods and the ways in which they are provided have also intensified. The growing corporate power of the retailers in the provision of food, together with the decreasing public confidence in many food products, has provided a major conundrum for the modern (post-1980) government. That is, how can government regulate food provision, given its longstanding priorities to 'deregulate' the economy but encourage the 'health of the nation'? While that particular question has under the Labour government been reformulated somewhat, so as to encompass the extent to which government can try to regain its regulatory legitimacy in relation to food, it remains highly pertinent. These questions take us to the heart of the state's capacity to regulate, especially at a time of crisis; to the interface between regulator and regulated; and to consider the ways in which both state regulation and retailer strategy may have changed. The very breadth of these issues means that we cannot hope to do them justice, but we do at least begin to provide material which indicates the trajectory of regulation in Britain.

Our interest in food therefore centres principally on the interface between political strategies for regulation, especially in relation to quality and hygiene, and the economic power of the major food retailers and their concern to promote food quality through the control of their supply chains. Essentially, it is in part the relationship between state and retailers that structures the choices of the foods that consumers buy (Flynn *et al*. 1998). We have been interested to find out the extent to which there may be a convergence or divergence between the practices and principles of economic and regulatory interests and in turn what this may mean for the representation of consumer interests.

The relationship between food, the consumer and choice is complex and multifaceted (see Murcott (1998) for a variety of approaches to this issue). We argue in this book that this cannot be fully comprehended in the British case without assessment of the role and position of the corporate food retailers. During the course of the past decade, food has risen to become one of the most prominent and sensitive public policy issues in Britain. In particular, questions about the safety and quality of food have spilled over from domestic policy arenas to the European Union (EU) and its other member states. The EU, in grappling with the demands of trying to create and manage a single market, with the obligations of the World Trade Organization and with the need to respond to new food technologies, has found itself drawn into an apparently never-ending stream of consultations and legislation on food. These issues in turn have had to be addressed by the member states.

Contemporary food policy, however, is not simply a matter of high politics (at the level of the superstate or nation state). Food also involves important social and economic considerations. For example, the British government, amongst others, explicitly recognised – though somewhat belatedly in the early 1990s – that there is a new link between food consumption and health (the 1991 Health of the Nation White Paper). The White Paper promoted the provision of quality food choices by 'informed consumers' in order to balance direct state action to ensure food standards with the encouragement of better food choices and options supplied by retailers. Meanwhile, food retailers across northern Europe are now major economic actors in their own right. Food is a growing economic sector in the EU and so will naturally attract considerable political attention. For instance, the number of retail shops in the EU is estimated to be almost four million, clustered particularly in the countries of Italy, Germany, Spain and the UK. The number of shops per 1,000 inhabitants is very high in countries with a strong tourist industry, such as Portugal, Greece and Italy, and here concentration is lower. The UK has the highest number of persons employed per shop and one of the lowest numbers of shops per 1,000 inhabitants. Shops in Greece and Italy employ the smallest number of persons. These country-by-country differences mask a more general process of concentration. For example, in 1994, the six major retail self-service outlets in each member state accounted for a significant market share in general food sales: 57 per cent in Germany, 61 per cent in Belgium, 35 per cent in Spain, 67 per cent in France, 39 per cent in Greece, 28 per cent in Italy. In Finland, the major retailing group accounts for 40 per cent of the market share, and the three big retailers account for 80 per cent. This growth in concentration is being increasingly matched by higher levels of cross-national activity, with the German firm Tengleman and the Dutch Ahold having over half of their turnover categorised as international trade.

The core of our thesis, which we develop here, concerns not just the changing nature of regulation (from public-interest to private-interest regulation) but also its focus (from emphasising production to emphasising consumption). Food is a particularly good policy area in which to explore these trends because it is one where there has been a clear shift reflecting economic and political power along the supply chain from agricultural production policy (see the work of Grant 1983; Cox *et al*. 1986; Smith 1990) to food manufacturers and latterly corporate retailers (see Marsden & Wrigley 1996; Marsden *et al*. 1998). One significant consequence for the corporate retailers is that their continued market power is increasingly dependent upon their social and political actions towards both state agencies and consumers. It is through these actions that market opportunities or 'spaces' can be kept open and legitimated on a continuous basis. As a result of this shift of emphasis, new power relations and new competitive relationships have emerged which implicate and influence the provision of foods; and indeed the provision of choices of foods the consumer is offered.

The changing roles of the state and of corporate retailers in delivering consumer-based rights, and the maintenance of regulatory conditions, are therefore far more embedded than hitherto in the consumption process (see Marsden and Arce 1995; Marsden and Wrigley 1995). These conceptualisations place an emphasis upon viewing retailing, the state and consumption as highly interactive agents in the dynamics of British society in the 1990s. In economic terms, we argue that the consumption nexus rather than the capital–labour nexus provides a more attractive location for the abstraction of profits and the development of capital accumulation (reflected in the increasingly different fortunes of primary food producers, manufacturers and the 'near market' agents, the retailers). And, that the combination of state and retailer relations and actions have increasingly become a more profound agent in the broader processes of economic and social restructuring in this modified sphere of consumption (see Chapter 3).

In the British case, at least, this means that retailers have become agents of social legitimation as well as exchange, tending to conflate questions of political citizenship with consumerism (see Chapter 4). The retailers' ability to continue to produce new food choices has occurred at a time when the systems of provision established from and by the state in the public interest have come under considerable pressure. Innovative and quality food choices have come from the corporate retailers, while the state, through the ideology of deregulation, has stimulated market competition and private-interest forms of regulation (see Chapter 7). This has not meant that actual deregulation has occurred. Rather, we have witnessed a process of re-regulation of food provision through the development of hybrid forms of state–retailer relationships and regulatory styles. The effects of these have been to sustain and expand the markets of the corporate retail sector and the legitimisation of neo-conservative policies based upon the continual extension of consumption rights (see also Saunders and Harris 1990). The extension of these consumption rights does not just occur through the selling off of former public assets (for example, through the 'right to buy' council houses, or the extension of share options); but it also has origins, we argue, in the development of a retailer-led food provision system.

The focus upon retailers, the state and food quality in this book thus provides an important window which has allowed, on the one hand, a more sensitive analysis of state regulation, and on the other, an understanding of the ways in which retailer competition and strategies for delivering quality food products are constructed. This can only be gained, we propose, by analytically linking the spheres of retailing, regulation and consumption. These spheres become the major focus of the book. Moreover, out of these interactions emerge the outcomes which dynamically construct the 'consumer interest'.

Food policy formulation and its implementation

In order to appreciate the dynamics of regulation and of the retail sector, we have sought to understand the making of policy and its outcome on the ground. To do so, we have followed the approach outlined by Winter (1990), which seeks to overcome a traditional divide which has bedevilled much implementation research. This is caused on the one hand by those who focus on why policy goals are rarely realised in practice (the so-called 'top-downers', for example, Pressman and Wildavsky 1973), but which tend to downplay the role of bureaucratic behaviour and the context in which it occurs; and on the other hand those who target the relationship between the regulator and regulated to understand policy implementation on the ground (the so-called 'bottom-uppers'), but who then often miss much of what happens to influence implementation that is beyond the world of the actors concerned. Winter (1990: 20) seeks to bring together 'the most promising theoretical elements' from existing work along with some under-researched variables 'to present a preliminary model to explain implementation outcomes'. According to Winter (1990: 20–1), implementation outcomes arise from four main sociopolitical processes or conditions: (1) 'the character of the policy formation process'; (2) 'organizational and interorganizational implementation behaviour; (3) street-level bureaucratic behaviour, and (4) the response by target groups and other changes in society'. As Winter points out, these processes also represent different levels in the policy formation and implementation processes. We reflect more fully on the policy formation and implementation process in the conclusions (Chapter 11), where we assess the prospects for the Food Standards Agency. At this stage, it is useful to make two points about Winter's model. First, we do not spend too much time on dealing with interorganisational issues, since central–local relations have been well studied and we have little to add here (see Rhodes (1997) for a comprehensive review). Second, the areas we do wish to focus on to deepen our understanding of implementation are *policy formation* and the *delivery of policy* (i.e. Stages 1 and 3). It is these two levels – simultaneously – which have been subject to most adjustment over the period, and which corporate retailers, through their organisational behaviour and strategies, have sought to influence.

Both Stages 1 and 3 of Winter's (1990) model have significant implications for the research process in that they involve gaining access to, and the trust of, key players. The details of the data collection method that we pursued are explained in the appendix. To gain an adequate understanding of the behaviour of Environmental Health Officers (EHOs) and Trading Standards Officers (TSOs) (our 'street-level bureaucrats'), we conducted research over twelve months in two contrasting field sites. One was an inner London borough (a unitary council which counts as a single food authority) and the other a county in southwest England (which consists of six food authorities, as EHOs are based at the district level). Given the sensitivity of the research, the food authorities concerned have been anonymised.

Contextualisation of the particular circumstances of food officials and the need to locate them within our broader understanding of the food policy process has been achieved by adopting an actor-oriented methodology. This has allowed us to explore the way in which actors in local situations are tied into wider sets of relations, and so helps to overcome the divide between the 'macro' and 'micro' (Murdoch and Marsden 1995: 368–70), an inevitable tension given the breadth of the study. Such an approach is employed to explore the relationship between policy implementation and outcome, and to embrace analytically the European, national and local levels of government. In so doing, we are able to show that policy and its regulation is not linear or step-by-step, but a highly contested and contingent process (as Clark (1992) suggests) which involves reinterpretation, transformation and reformulation during the implementation process by a whole range of different actors caught up in food retailing, regulation and consumption.

Actors, moreover, are located in concrete situations or 'actor spaces', for example, the retail boardroom, outlet or the local enforcement office. When we consider the relationships between the local and the national, we therefore do so from the position of these spaces (see Murdoch and Marsden 1995 and Chapters 8–10). We can understand structure and structural outcomes, and their (re)production, by investigating the microprocesses through which actors interact, and through which the local and the national are 'mixed up'. We do this, methodologically, where we follow the actors as they build these associations and interact, negotiate and struggle with each other. Throughout the book we make extensive use of the data that we have collected, so allowing the actors to 'speak'. In Part II we use detailed interviews with strategic actors in the retailing, regulation and consumption 'matrix'. In Part III in particular, we explore primarily the links between the actor spaces of the food regulatory framework (within our two study areas) with ethnographic evidence, presented through interview extracts; but secondary materials documenting the more formal interactions are also used to illustrate the process by which specific actors are tied into wider sets of relations. For example, by exploring the linkages between actor spaces within the food regulatory framework, we are also able to illustrate the 'living out' of the macro model of private-interest regulation that we develop in Chapter 7.

Perspectives on the consumer, retailing and regulation

Our concern with local processes of implementation and with local food regulators on the ground has not been at the expense of the policy formulation process. As we made clear above in following the model outlined by Winter (1990), we have sought to link together both formulation and implementation in a non-linear or directional fashion, and to show the ways in which implementation practices feedback through to the reformulation of policy over time. More generally, this policy process provides a window on the restructuring of the state and its means of regulation. However, it is not simply a case of the state determining or even directing regulatory arrangements; rather, it involves a much more complex set of relationships with food retailers and local government officials. These relationships take place along the policy process and at different scales and levels of government.

The book is organised into three sections to allow us to explore these interactions between the triumvirate of consuming interests – consumption, regulation and retailing – at different spatial scales and levels of abstraction. Thus, each section of the book tackles these themes. Part I provides a policy and conceptual framework which places into context the changing nature of food regulation and retailing. At a conceptual level, food consumption and regulation are linked to citizenship and notions of freedom. Part II grounds our conceptual and policy framework into national arenas and shows how the nation state remains an important site for regulatory activities and the construction and maintenance of retailers' 'competitive space'. Meanwhile, the marginal impact of the consumer is analysed through the operation of the consumer lobby. Part III provides our analysis, based upon detailed local-level work, of the operation of the food regulation system. Here we are able to show how consumers, regulators and retailers operate within their own spaces and how those interact with, or pass by, one another.

We start in Chapter 2, paralleling Stage 1 of Winter's (1990) model of implementation, providing an introduction to the ways in which the British government and, increasingly, the EU shape food policy. This draws out the organisational base of food policy and the tensions between different actors. While the consumer may have moved to the centre of policy debates, policy itself remains driven by economic forces. Regulation is operating increasingly in tandem with the supply chain regulation of the major retailers.

Central to the reasons why consumer organisations have played a marginal role in food policy debates compared to retailers is that retailers are often willing and able to present themselves as those best able to represent and therefore to 'construct' the consumer. Increasingly, consuming interests are taken up and articulated in relation to the supply chain itself, and this tends to give particular opportunities to the ascendant corporate retailers. Positioned as they are at the point of sale, corporate retailers can begin to represent the consumer in ways which enhance the retailers' role in the regulatory supply chain itself (Chapter 3). As we document in Chapter 4, the rise to prominence of retail capital is associated with a reformulation of citizens' rights to food. Broadly, it is possible to distinguish between a *freedom from* want and a *freedom to* consume. It is the latter which is of greatest interest to retailers, and manufacturers, as it permits them to construct differential notions of food quality and then to allow consumers to make their 'choice'. The organisation of consumer interests is analysed in Chapter 5. The important role that government has traditionally played, and continues to play particularly at a European level, in representing the consumer is highlighted. The marginalisation of consumer groups is explored in relation to their structure and links to government.

Retailers, in contrast, have become increasingly dependent for their economic success on their regulatory embeddedness (Chapter 3). It has been important for retailers to demonstrate their customer credentials to government and regulators, to project themselves as the new custodians of the food provision system during a period when other parts of the up-stream food sector have been the locus of public concern. A main innovation in food quality enhancement has been in the development of retailer-led food hygiene and hazard systems. These have increasingly been developed as a condition of market entry for food suppliers and manufacturers. Hence, as far as the supply chains are concerned, it is increasingly not enough to supply quality foods of the right compositional standards. It is also necessary to demonstrate that systems of quality management have been put in place (such as Hazard Analysis and Critical Control Point systems (HACCP)) as a food assurance scheme. Hence the retailers expect more and more from their suppliers in terms of the policing of food delivery as well as the type specifications of the food produced. This stands to give retailers a market advantage with customers, and it demonstrates to government that they are taking existing food regulation seriously (particularly the 1990 Food Safety Act and the 1993 Food Hygiene Directive).

However, these systems of quality construction are seen as a moving target and hold important implications for the maintenance of the competitive spaces between the big retailers. They are the tangible ways in which the consumer interest is constructed on a dynamic basis. Retailers are at the apex of this quality construction; being able to absorb and transmit regulatory changes, customer reactions and supply chain quality assurance parameters. All this is a highly contingent and differentiated process, as we demonstrate in Chapter 7, depending increasingly upon the size and power of retailers' competitive space between themselves and *vis-à-vis* others. For the customer this means that food quality is both highly differentiated above the state-defined baselines, and that the choice of retail outlet will affect the type and the constructions of quality purchased and consumed. In this sense, the mass activity of retail shopping resembles the act of choosing a restaurant for a meal out. The choice of location will engender different quality assurances and constructions that are consumed.

Critical to the maintenance of retailer competitive space and the freedom to act in the market in such ways as to foster their dominance is the participation of food retailers in regulatory activity at the local, national and (increasingly) international levels. In Chapters 3 and 6 we identify the regulatory domains in which retailers operate. Overall, retailers make a considerable contribution to the co-evolution of both a regulatory culture and a consumer culture based upon governance models which prioritise the facilitation of market opportunities for retailers. This also has the effect of reducing the powers of articulation of other representative bodies.

In Chapter 7, we present a framework by which we can begin to interpret the changes that have occurred in food regulation during the 1990s. We identify the emergence of a private-interest style of regulation in the food system that contrasts with the traditional regulatory style based upon notions of the public interest. In the latter case, it is central government that sets standards through legislation which are enforced locally, on behalf of the public by officials such as EHOs, TSOs and Public Analysts. They ensure similar baseline standards for all consumers. In its more traditional form, food regulation has sought to ensure a combination of security of supply, accessibility, affordability and safety. As we show, however, the major food retailers, conversely, voluntarily regulate their own systems at their own expense, promoting individual choice based on their own hierarchy of quality definitions. So as the major retailers have become the principal actors in the food system, they have negotiated key responsibilities in the management and policing of that system.

Releasing powers of regulation has not meant that the traditional public sector role of protecting the consumer from health risks associated with food has been subverted. Rather, what has happened is that different sets of rights, intimately linked to private sector provision, have been fostered. As we argue in Chapter 3, consumers are empowered, and are free to make the choices as they see fit. These are competing notions of rights, and are associated with different regulatory arrangements. The public or the private sector may take the lead as appropriate, but in practice they are to be found alongside one another, increasingly upholding a bifurcated food retailing structure and regulating system. This is, however, anything but a smooth process. The resulting tensions within the regulatory framework have been particularly acute at the local level of implementation (which we explore in Chapters 9 and 10). What we are witnessing, therefore, is the government essentially trying to act as backstop, to ensure basic standards of food safety. Over and above this the multiple retail outlets are creating additional rights based on their different guarantees of food quality, which are available to their customers. This requires these firms to police and regulate their own food chains.

The implications for the representation of the consumer interest are significant. For example, on food irradiation, one Ministry of Agriculture, Fisheries and Food (MAFF) official argued that on scientific grounds there was no reason for not approving the process. 'But', he continued, 'the fact of the matter is that it is simply not being used in this country because the retailers have said they are not going to handle it. Consumers will take fright, you see.' This statement is important, not only because it recognises that retailers are regarded by government as able to represent consumers, but also because it means that government is increasingly engaged in sharing its authority for food safety with the major retailers. Here, then, is a classic expression of governance, a broadening of government from the nation state to embrace private interests. The linking of the public and private sectors to deliver policy goals is a central thread running through the book, and a clear indication of the broader nature of the restructuring of the state that is currently underway.

In Part III, we explore at the local level the articulation of these state–retailer relationships, and in particular seek to progress our understanding of the private-interest model of regulation. Thus Chapter 8 charts the ways in which the structural changes that have taken place in food retailing at the national level have impacted in our local study areas. This provides a picture of the nature of food retailing that EHOs and TSOs have to regulate. The chapter takes forward a retail-geography perspective by analysing not only the changing spatial structure of retailing but also the nature of the relationship between retailers and the local state and the relative ability of the latter to influence the local geography of corporate restructuring.

Having presented a picture of local retail spaces, Chapter 9 charts the efforts of government and key retailers to nationalise the food regulatory system in the 1990s. This chapter provides an in-depth study of the dynamic interactions between local and national regulatory practices, and is able to depict a regulatory advance, a retreat and a rapprochement. The outcome is the increasing differentiation of regulation between the tiers of retailing, and thus the substantiation of a key element of the model of private-interest regulation that we presented in Chapter 7.

Despite these moves towards the nationalisation of food regulation, however, local variation in enforcement practices has persisted, as documented in Chapter 10. In order to understand the geography of regulation, that is, the variability in enforcement, we address the administrative context in which EHOs operate and assess how this influences their implementation practices. That regulators do not always deal with those they regulate in a uniform manner is part of their coping strategy – a well-known phenomenon in implementation studies – but has important implications for the government's efforts to raise food hygiene standards. This is one of the points that we develop further in the conclusions to this book.

In conclusion, (a) we make an assessment of the retailers, regulation and consumption interactions depicted in the analysis, and (b) we do this in relation to the interactive relations and dialectics which exist between strategic food policy formation and its implementation. In doing this it is important, we argue, to consider the social and political location of corporate retailers in the UK in the 1990s and to assess how they are able to sustain a retailer-led model of food governance in a period of growing, if uneven, consumer consciousness. To understand these relationships, it is necessary to bring together the policy studies literature with that of the cultural and regulatory geography of retailing, because the construction of the consumer interest, and more generally the social provision of food choices, transcends economic, social and political spheres. Corporate retailing, regulation and consumption are thus welded together in the current British model of retailer-led food governance; and their particular configuration is becoming a major feature of modern British food consumer culture and polity.

Part I

Concepts and framework

2 Food policy and regulation

Introduction

In this chapter, we will outline the ways in which the British government, and more recently the European Union, have shaped postwar food policy. As the previous chapter outlined, our argument is that food regulation, retailing and consumption have become increasingly significant focal points of analysis. Here, we link those debates to the development of contemporary food policy and show how the policy has shifted over the postwar period, from one linked a priori to a productivist agriculture to one based a priori around consumption-oriented retailing, an argument which is developed in the next chapter. The growing involvement of the European Union in British food policy, particularly in relation to food safety, has not undermined the trajectory of the development of British policy. Instead, there has been much common ground between British and European Union food policy makers in the 1980s and 1990s; for example, in the role of regulation in promoting trade and in the increasing emphasis on self-regulation. What has tended to happen is that on the one hand, the gathering process of Europeanisation of food accentuated trends in the 'techniques' of regulation within Britain, while on the other hand, it raised consumer and social issues which had hitherto been marginalised during the period of Conservative government (1979–97).

Perhaps most obviously, the European Union has raised the *profile* of food issues. This has happened in two ways. First, as an indirect effect of the system of agricultural support in Europe, the costs and surpluses were more apparent to consumers than under Britain's former system of deficiency payments. Second, efforts to create a 'freer' market in food products in the 1980s, as part of the overall European project of developing a Single European Market, raised questions as to what counted as a particular food product, raising cultural and national sensitivities to food that had previously lain dormant.

The chapter provides a thematic analysis of the development of food policy. At its centre is the role played by the state as the remaining and ultimate arbiter of rules. The role of government (at both a national and European level) is a dynamic one, and in conjunction with other interests, particularly economic ones, it helps to structure the food choices that consumers make. In comparison to economic interests such as food manufacturers and retailers, consumer groups play a more marginal role in the development of food policy. The reasons behind the relative weakness of consumer organisations and the representation of the consumer are more fully covered in Chapter 4, which explores the links between food rights and citizenship, and Chapter 5, which examines the organisation of consumer interests at British and European level. Here it is argued that producer interests have long been at the heart of food policy making. At first it was farmers who dominated. During the early postwar years, food policy was subservient to agriculture policy, a point highlighted by the merger as a junior partner of the Ministry of Food with the Ministry of Agriculture. (The subservience of the consumer to producer interests and its implications for food choice is explored in Flynn *et al.* (1998).) Below, we outline some of the key links between economic interests and government and how they can shape British food policy. The opportunities for national governments to determine their own food policies are, however, increasingly limited, and so the chapter also documents the development of EU food policy, the challenges it faces and where the responsibilities lie for different regulatory strategies.

Economic interests, government and food policy in the UK

The high point of farming influence on government policy in Britain is now long past. The protracted relative economic and political decline of farming has been accentuated by the dramatic rise to prominence of first food manufacturers (Flynn *et al.* 1991), and more recently retailers (Wrigley 1991, 1992, 1994), which have proved to be the two most buoyant sectors within the food system. Throughout the 1980s and into the 1990s, as we explain in the following chapter and Chapter 6, the major retailers underwent considerable expansion such that today they have captured about two-thirds of food retail sales. In contrast, the smaller independent retailers account for an ever-declining proportion of sales. Instead of farmers, it is actually the major multiples that increasingly determine the shape of the British food sector and are able to influence the food choices on offer. Together, retailers and manufacturers have been sources of considerable innovation across a range of areas, from new products to the distribution and storage of those products.

Within the food system, it is the manufacturers and increasingly the major retailers who are exercising influence over farmers (Flynn and Marsden 1992; Marsden and Wrigley 1995) and help to ensure the continued marginalisation of consumer interests. Such influence, however, carries with it a regulatory dimension. Retailers in particular have found themselves both drawn into and actively seeking a regulatory role. For example, during the late 1980s and into the 1990s, it was by no means apparent that key retailers would move into the dominant position within the food system they enjoy today. Retailers faced intense internal competition and growing criticisms of their sourcing strategies, and as a result had to grapple with the challenge of market maintenance (Marsden *et al.* 1997). This involved them in a more diversified set of relationships with the state at both the national and European level. Perhaps fortuitously, at the same time MAFF (the Ministry of Agriculture, Fisheries and Food), which retained considerable regulatory responsibility throughout the food chain, had been seeking to share some of its regulatory burden, often under the mantle of deregulation (see Chapter 7). That such a coincidence of regulatory interests could be realised owed much to MAFF's knowledge of, and involvement with, the industry built up over long years of its support. As one industry interviewee put it: 'the food industry has always had a good working relationship with MAFF because MAFF needs such a relationship. You see, MAFF is sponsoring the food industry.'

Contacts between retailers and government occur on a regular basis and at a variety of levels from the highest circles of policy making to local-level policy implementation. Such links owe much to the Ministry's preferred administrative style, which is to incorporate favoured (i.e. economic) interests in the decision-making structure in an attempt to ensure that the two move together.

Until recently, the non-party political nature of food policy has meant that the area has for the most part been regarded as a somewhat technical field. Decisions and advice on policy were best left in the hands of 'experts', thereby effectively excluding consumers and privileging the specialist knowledge of the food industry. Much work has therefore taken place in the Ministry's specialist advisory committees, notably the Food Advisory Committee, and in the Department of Health's Committee on the Medical Aspects of Food Policy (COMA). These Committees cover such topics as nutrition, food contamination, labelling, food composition and health and diet (see Foreman 1989: 56). The Committees exist to help protect the consumer, but they also assist in the coordination of government and industry activities (National Consumer Council 1988), and have been subject to increasing criticism for their lack of independence, unrepresentative membership and secrecy of proceedings. When in 1993 the financial links between members of the Food Advisory Committee and food companies were published for the first time, 'Twelve out of seventeen members … declared some cash reward', and that may have been an under-reporting (*Guardian*, 12 May 1993).

Today, the deliberations of food-related committees and their recommendations are significant for food choice because they are able to propose modifications to the boundaries of existing regulations. As the food companies and retailers search for new products and processes there is frequently a need for new regulations or the amendment of existing regulations. A good example of the types of changes that may be made are reflected in MAFF's attitude to food regulations which have 'tended to move away from imposing compositional standards on food ... to provide more effective ingredient and nutritional labelling' (Foreman 1989: 118). The extra flexibility this modified stance has given food manufacturers is reflected, for example, in low-fat products which were previously outlawed by specifications on minimum fat content. While Foreman is probably accurate in her characterisation, it is somewhat disingenuous to portray MAFF as an independent actor. MAFF was able to change British food law in this area because that was the direction in which European legislation was moving (for example, Directive 79/112 on food labelling (Council Directive 79/112/EEC, OJ no. L 33, 8.2.79, p. 1)). Today virtually all British food law is a response to European food law (MAFF Food Safety Directorate 1996) and it is to the European dimension of food policy that we now turn.

European food policy

Policy development

Food, as one of the key economic sectors within the EU, has long been a target of policy makers. At the European level, food policy has largely been shaped by the combination of two factors. One is the Common Agricultural Policy, which has led to an emphasis on product quality specifications. The other is the implementation of the internal market, which has meant in particular measures to ensure the free movement of goods (Commission of the European Communities 1997: 5). The result, however, has often been confusion over the basis of regulation and its objectives; or as the Commission openly admits, complexity, fragmentation and incoherence (Commission of the European Communities 1997: 12). Here we explore the initial ideas and principles surrounding EU food policy and the key factors which led to a rethinking of its approach. This is followed by a discussion of the organisational base of policy, for it is the roles and perspectives of EU DG (Directorate General) III (internal market and industrial affairs) and DG VI (agriculture) which have done so much to nurture and sustain competing perspectives on food policy and regulation.

Efforts to secure harmonisation of foodstuffs legislation, in order to allow the free movements of goods, began in the early 1960s and continued until the mid-1980s. This meant laying down specifications for different products (such as preserves, jam and chocolate). This so-called 'recipe law' approach, known more technically as vertical harmonisation, was not particularly successful. It was inevitable that Community-derived concepts of 'good beer, good sausages and good bread' should conflict with beliefs and sentiments as diverse as those constituted by the member states, which each had their own culinary traditions. Moreover, the diversity of climate and agriculture throughout the Community meant that the nutritional needs of the population were met in a variety of ways. Even in areas having access to the same raw materials, methods of preparation of food varied widely (Gray 1993: 1).

The national laws of member states had not developed from a sound logical approach to the need to create a consistent set of rules for behaviour between manufacturers, traders and consumers, but rather from their idiosyncratic and often ancient traditions. The harmonisation procedures therefore led to conflict between these culinary cultures and traditions, as the Community attempted to unify products which had culinary diversity into unique product prescriptions. These early attempts to legislate based on the harmonisation of produce specifications – the so-called vertical legislation – thus met with little success since they were perceived as an assault on long hallowed traditions.

During the 1970s, the production of a set of rules that could apply throughout the Community had begun to seem an impossible task, not least because agreement was rarely achieved as to which rules should remain or which could be removed (Perissich 1993). The result of this disagreement was a logjam of proposals within the Community's bureaucracy, as no country wished, as Gray (1990: 112) comments, to sacrifice its culinary culture on the altar of uniformity. While, by common accord, the member states of the European Community had agreed to set up a system of law which would transcend national frontiers, the deficiencies of the vertical approach necessitated a reappraisal of the role of food law in public policy.

The Commission's efforts to harmonise food law were effectively undermined in a celebrated ruling of the European Court of Justice in Luxembourg in 1979, the 'Cassis de Dijon' judgement (*Rewe-Zentral AG* v. *Bundesmonopolverwaltung für Branntwein* (Case no. 120/78) [1979] ECR 649), which presaged the abandonment of the contentious 'recipe laws'. Instead, the Court defined two new approaches to the Community's food legislation system. These were, first, the concept of *mutual recognition* (based on Articles 30–6 of the EEC Treaty), which attempted to overcome the inter-state conflict that the 'recipe laws' had instigated (Perissich 1993); and, second, the *principle of proportionality* which reduced considerably the number and the powers of those food laws in preparation and led to a restatement of the aims of food regulation within the Community as a whole (Commission of the European Communities 1985).

Broadly, the Court's ruling set the 'common interest' of the Community citizens against the interests of the system of regulation that demanded a unique definition of, in the case of the Cassis de Dijon, the alcoholic composition of all liquors. The Court considered there were, instead of the 'recipe law' approach that the Community was attempting to employ, various possible alternatives in this process of harmonisation. The one chosen, the Court believed, should be first, that which achieved the desired public policy ends and the protection of the consumer; and second, should not at the same time unduly restrict trade. Importantly, the Court judgement implied that the member states should no longer be expected to rescind existing national food laws for the purpose of harmonisation within the Community as the 'recipe law' approach had dictated; rather, they should instead first be expected to adapt their own rules to Community principles and, second, be expected to mutually recognise the laws of other member states. As Perissich (1993: 58–60) explains, this concept of mutual recognition meant that member state A could no longer subject a product to a control merely because it came from member state B, even if it was of the opinion that official control in member state B was inadequate. While this new procedure overcame the conflicts consequent upon the harmonisation tactics of the 'recipe laws' it did, as Perissich notes, create others as it necessitated a radical change in thinking for member states of an insular nature that cherished a historically conditional distrust of all things foreign.

In addition to this, the Court decided that in terms of public policy in the food sector it was also necessary to operate a principle of proportionality. In terms of the application of food regulation, this principle implied that the 'essential' should be separated from the 'optional'. It also encouraged the application of legal measures to foodstuffs only when 'genuinely necessary' to achieve the desired objective (Perissich 1993). Consequent to the Cassis de Dijon judgement, therefore, the Community has been able to define a system of food legislation in which it is able to more reasonably justify that it is only concerned with provisions that it considers are justified as being in the general interest, namely the protection of public health and consumers (Commission of the European Communities 1997: vi).

The slow and difficult progress on harmonising food products, the Court judgement that such a process of harmonisation was unnecessary, and preparations for the Single European Act in 1985 all hastened a rethink of Commission strategy. In a highly significant communication of November 1985, 'Community legislation on foodstuffs' (Commission of the European Communities 1985), the Commission commented upon the difficulties encountered in the implementation of Community-wide food legislation, and further, stated that it was 'neither possible nor desirable to confine in a legislative strait-jacket the riches of the twelve European Countries'. At one stroke the Commission recognised the different traditions of the member states and the different compositions of products. The Commission's proposals, therefore, reinforced the principle of proportionality as a legislative procedure which distinguished between 'on the one hand, matters which by their nature must continue to be the subject of legislation and, on the other hand, those whose characteristics are such that they do not need to be regulated'. Most importantly, for the first time the Commission actually defined the limited basis upon which such legislation on foodstuffs could be justified, thereby defining the responsibilities of the Community's regulatory food system as a whole:

> Community foodstuff legislation must ensure that there is a high level of public health protection and that the consumer is accurately and adequately informed as to the nature, characteristics and, where appropriate, the origin of foodstuffs placed on the market.
>
> (Commission of the European Communities 1985)

In effect, the consumer becomes the centre of regulation, as later Commission documents have confirmed (see, for example Commission of the European Communities 1997: vi). But it is important to remember that the maintenance of the internal market is also very significant. Indeed, it is the single market which has directed where the Commission takes action in relation to the consumer.

According to a Commission official, this has meant that the Commission have concentrated on three areas:

1   public health (e.g. additives)
2   protection of the consumer interest (e.g. labelling)
3   control and enforcement provisions (e.g. hazard analysis)

It is on the third of these that we wish to concentrate here, through a discussion of the adoption of hazard analysis at the European level (the role of hazard analysis at a national level is covered in Chapter 6 and at the local level in Part III). Before embarking on an analysis of regulatory strategies, however, we outline the key responsibilities for food policy making at a European level.

Responsibilities for food policy making

Within the EU, there is a basic division of responsibility for food regulation between DG III (responsible for the internal market and industrial affairs) and DG VI (agriculture). This division is enshrined within Annex II of the Treaty of Rome, and allocates responsibility for primary agricultural products to DG VI and for non-Annex II products to DG III. Commission officials are keen to stress the level of cooperation that exists between the two DGs, with regulation carried out by mutual agreement. For example, general labelling and additives rules will be produced by DG III and will apply to all foods unless there are specific reasons for their exclusion. Nevertheless, the two DGs have been alternative sources of authority for food legislation, and have in the past adopted distinct regulatory strategies (as explained below). More recent organisational changes to take account of the political response to growing consumer worries about the safety and quality of food are covered in Chapter 5.

DG VI hygiene rules normally stop at the wholesale stage. This approach is to put responsibility for hygiene on each economic actor up to that stage. The obligation on the primary producer, however, is minimal (for example, it is at the slaughterhouse where meat is inspected and not at the farm). DG III legislation applies to all foods and covers all stages of production, but excludes primary production. Thus the retail sale of meat products is covered by DG VI, but at the point of sale to the consumer DG III legislation (the Food Hygiene Directive) will prevail (see below).

In a similar way to Britain, committees of technical experts have an important role to play in advising officials and senior political figures. Amongst the most important committees operating at a European level are the Scientific Committee for Food, which provides advice on any problem relating to the protection of the health and safety of people in connection with the consumption of food (with members drawn from the academic scientific community); the Advisory Committee on Foodstuffs, which may be consulted by the Commission on the harmonisation of legislation relating to foodstuffs (with membership of this Committee including two consumer group representatives, but otherwise largely drawn from industry); and the Standing Committee on Foodstuffs (SCF), which provides opinions on matters under consideration by the Commission (a permanent committee of government officials).

Again, like Britain, there have been concerns raised as to whether the SCF is fully considering consumer concerns (McColl 1992). The Committee consists of seventeen scientists drawn from a variety of disciplines concerned with food safety, who serve for a renewable term of three years. But while its expertise has been concentrated on toxicology, biochemistry and nutrition, food issues on the agenda in the 1990s require a wider scope of academic expertise; for example, on microbiology, novel foodstuffs, methods of analysis and risk assessment techniques. In addition, the SCF has been criticised for being unable to respond quickly to the growing number of food issues under consideration (Consumers in the European Community Group 1992).

Of more importance perhaps to the efficacy of the regulatory system and the protection of the consumer is the selection of experts, and interests represented, on the Committee itself. The Commission has been criticised for its increasing reliance upon Committee members, who are essentially chosen by the Commission rather than by autonomous means (Chambers 1990); further, consumers are not represented on the SCF at all. The Commission has taken the view that this would be inappropriate for an expert committee. However, the work of the SCF involves decisions on risk assessment and risk acceptability which incorporate societal value judgement. It seems reasonable, therefore, to expect some opportunity for consumer participation in this decision-making process (McColl 1992).

Strategies for regulation

DG VI has been highly prescriptive and provided detailed qualitative standards for different products. In other words, there is a large prescriptive element to regulation which details the ways in which results are to be achieved. In contrast, DG III, where possible, favours rules which fix an obligation to achieve results and leave it to the competent authority to decide how the results are to be achieved.

Increasingly, however, officials recognise a convergence in thinking between DG III and DG VI. Where possible, both directorates are seeking to apply hazard analysis to the formulation and revision of regulations.

Hazard analysis is increasingly widely applied in a number of regulatory fields, such as the environment. Within the area of food policy, it was pioneered by the international regulatory body Codex Alimentarius. From the perspective of policy makers keen to break down trade barriers, it provides an attractive approach. This is because it seeks harmonisation through results, rather than means, and thus avoids the need for detailed harmonisation legislation. Within the EU, hazard analysis, or as it is commonly known, HACCP (Hazard Analysis Critical Control Point System), found a receptive audience. Amongst policy makers, there was increasing recognition that, first, EU food legislation was deficient (e.g. rules were constantly behind technical developments, enforcement was highly variable), and second, that the existing approach may not have been particularly effective at protecting the consumer from unhealthy food (inspectors are partly dependent on the quality of their checklist).

The most significant contribution of hazard analysis is in the Food Hygiene Directive (Council Directive 93/43/EEC, OJ no. L 175, 19.7.93, pp. 1–11) which lays down general rules of hygiene and procedures for verification of compliance with these rules (i.e. HACCP). The Food Hygiene Directive is in turn a daughter directive of that on the Official Control of Foodstuffs (Council Directive 89/397/EEC, OJ no. L 186, 30.6.89, p. 23). This was an important framework directive for encouraging the harmonisation of food law between member states, which was later supplemented by the Additional Food Control Measure Directive (Council Directive 93/99/EEC, OJ no. L 290, 24.11.93, p. 14). It was intended to achieve a consistent approach to food law enforcement by laying down a number of general principles:

- food should be inspected regularly at the point of production to avoid the need for border controls between member states;
- inspection should be harmonised between member states;
- there should be mutual recognition of standards within the European Community;
- details of member states' food law enforcement programmes should be submitted annually to the European Commission (MAFF 1996: 46).

The Food Hygiene Directive covers the preparation, processing, manufacturing, packaging, storing, transportation, distribution, handling and offering for sale or supply of foodstuffs not covered elsewhere by product-specific hygiene Directives. It also covers gaps in the product-specific hygiene Directives. For example, few of these Directives cover the sale or supply of their foodstuffs, and where they do not, then Council Directive 93/43/EEC does (MAFF 1996: 50).

Within the Directive Article 3(2), there is the demand that:

Food business operators shall identify any step in their activities which is critical to ensuring food safety and ensure that adequate safety procedures are identified, implemented, maintained and reviewed on the basis of the following principles, used to develop the system of HACCP (Hazard analysis and critical control points):

- analysing the potential food hazards in a food business operation;
- identifying the points in those operations where food hazards may occur;
- deciding which of the points identified are critical to food safety – the 'critical points';
- identifying and implementing effective control and monitoring procedures at those critical points, and
- reviewing the analysis of food hazards, the critical control points and the control and monitoring procedures periodically and whenever the food business operations change.

Although it would have been possible to add two additional requirements (for example, the need for documentation on cleaning procedures which consumer groups would like to have seen, and verification procedures) to fully bring the Directive into line with the Codex Alimentarius, the formulation of HACCP nevertheless marks a highly significant step in that direction. The Commission has defended its position on the grounds that 'it was not considered necessary to lay down formal requirements regarding verification and documentation. Each food business is left with the flexibility to decide what requirements are necessary, subject to the supervision of the Competent Authority' (Commission of the European Communities 1997: 29). As we shall see (in Chapter 7), the ability of retailers to engage in hazard analysis and manage risks along their supply chains is highly variable and an important source of differentiation in both their ability to regulate themselves and of the attitude of environmental health officers to them.

Conclusions

Two points are significant in considering the Food Hygiene Directive, the implementation of which at the local level is explored in Part III. First, the Directive works with the grain of the regulatory practices of the major retailers and food manufacturers. As one senior official at the Department of Health, whose sentiments would be shared at the European level, remarked, 'yes, I think the large retailers are very much capable of regulating themselves'. The emergence of self-regulation in which HACCP plays such a key role has, however, been fraught with difficulties. The EU's early approach to the regulation of food became bogged down in a mire of detail, which was not helped by competing perspectives from DGs III and VI. Second, there are clearly important parallels between British and European regulatory strategies.

Within both tiers of government, there was a growing dissatisfaction with a reliance upon compositional standards to regulate food. Both recognised the value of alternative forms of regulation based upon hazard analysis, to regulate the process as opposed to the specific content of foods. Commission officials have some sympathy with the claims of the then Conservative government that the UK had been keen to promote the Food Hygiene Directive. However, Commission officials would also want to draw a distinction between a simplification of regulation and a lowering of standards, and would recognise the need for government to play an active role in the control and enforcement of food standards.

A third and more general point to emerge is that national forms of food regulation have increasingly been superseded by legislation decided at the European level. As the Commission notes: 'Today, the vast majority of food law has been harmonised at the Community level, and in many fields the scope for unilateral action by the Member States is severely restricted' (Commission of the European Communities 1997: 13). The Europeanisation of food policy making clearly marks a significant shift in the regulatory domain in which British retailers and regulators have to operate. Tensions between competing sources of authority and strategies for food policy making are creating opportunities which economic interests, particularly retailers, are seeking to fill. Both regulatory and economic interests are now coalescing around the notions of self-regulation and hazard analysis. In the next chapter, we examine the ways in which the restructuring of British retailing is linked to these processes of regulatory change. While in Britain, food policy making has tended to lose its productivist priorities, this has been matched by the steady growth of a European influence which encourages private sector responsibility over and above complex compositional control of the quality of foods. Located within these national and European shifts in policy styles have been the emergent British corporate retailers. To survive and prosper, corporate retailers need to create and maintain local and national embeddedness; and here the nation state remains a key site for the resolution of the competitive imperative of retailers.

3 Restructuring and retailing

Introduction

At the experiential level, most of the public take it for granted that large retail outlets are at the vanguard of food supply and purchase. In the academic and public policy literature, however, there is a significant paucity of work which portrays the social and political significance of these retailers. Agrarian political economists have had some difficulty in locating corporate retailing into their broader, and still largely production-oriented, models of global 'food regimes' (Goodman and Watts 1994; Le Heron and Roche 1995). Similarly, rural sociologists and retail geographers have tended to treat retailing as a discrete economic sector, somewhat unleashed from the slower-moving food production and manufacturing sectors. In this chapter, we build upon our earlier analysis of food policy and regulation and begin to assess the significance of corporate retailing in a social and political sense.

We will give particular emphasis to considering evidence of retail power; how, through social and political means, economic and market dominance seems to be maintained. As Christopherson (1993: 274) argues, the rules governing investment within and competition between firms 'constitute environments for capital accumulation', and produce 'quite different patterns of economic behaviour within and across national boundaries'. We argue that this expression of economic uneven development in retailing has to be socially and politically constructed in separate regulatory arenas operating at different spatial scales. Feeding these are highly contested knowledges about food quality and constructions of the consumer interest.

Thus far, most attention has been given to the local retail contestations, or 'store wars', associated with retailing and the planning system in the UK. But as Wrigley and Lowe argue:

> Orthodox retail geography has been remarkably silent about regulation and the complex and contradictory relations of retail capital with the regulatory state. With the exception of a rather one-dimensional discussion concerning the constraining influence of land-use planning regulation and some limited debate about shop opening hours regulation, the

transformation of retail capital appeared to take place in a world devoid of a macro-regulatory environment shaping competition between firms, the governance of investment, the use of labour and the overall extraction of profits from what Appadurai calls the 'situation of exchange'.

(Wrigley and Lowe 1986: 13)

In addition Pred (1996: 14), in reviewing consumption identity and the practices of power relations, has commented on the 'all too thin literature on consumption and the political economy of regulation and deregulation'.

The intense competitive environment in which corporate retailers are located in Britain means that they are forced to participate and to attempt to shape a whole series of regulatory domains; such as land-use planning, environmental regulation and food law. The corporate retailers cannot accept any notion of 'saturation' over the demand for retail goods (Langston *et al.* 1995). Rather, they have to become involved in the shaping of competitive spaces which 'prise open' new territories and terrains of exchange and profit making.

In this chapter, we outline some of the recent (post-1990) trends in the management of the British retailers' competitive space. In doing this, we argue that this continuing British 'success story' is based upon a growing regulatory 'embeddedness' on the part of the retailers, suggesting the reformulation of traditional conceptions of economic and state power. Later in the chapter, we explore some of the implications of these developments for our analysis of food and state action in Britain. The reasons for undertaking these re-conceptualisations concern the increasing need to understand how and why sections of society take opportunities and make choices; and particularly how combinations of public and private actors construct value around food products and provision.

Retailers' regulatory terrain: maintaining uneven competitive space in Britain

The provision of food is undergoing a process of global transformation, and it would seem that this is a significant indicator of more underlying value changes which are occurring, associated with globalisation on the one hand and the reformed nature of uneven development on the other. Unlike in other sectors, however, such as manufacturing, these global trends are not necessarily built upon the global interconnectedness of retail capital. While there is evidence of retail buying alliances and some cross-national activity, the key to globalised forms of retailing lies in the potential for convergence of national and international regulatory conditions. While these may be principally seen to be the preserve of the nation state, retailers, through their international and national political and regulatory activities, are increasingly influencing the reconstitution of those states. Of particular significance here is the increased national flexibility of retail concentration. The ten largest chains in both the EU and USA account for roughly 30 per cent of total grocery sales; in individual

US states and EU member states (most notably, Denmark, the Netherlands, France and the UK), the three largest retail chains account for between 40 and 60 per cent of the grocery market. This not only gives them a strong national base from which to explore transnational expansion, but it also enhances their powers over the food processing and wholesaling sectors, which increasingly have to answer to the retailers on quality and value grounds.

In the British case, the rising economic and market significance of the corporate retailers is well recognised (see Marsden and Wrigley 1995; Flynn *et al.* 1994). While global pre-tax margins are relatively low (around 0.5–3 per cent), with gross margins ranging from 20–30 per cent, improvements in efficiency, reductions in costs of goods sold and increases in the turnover of sales can each have a potentially large influence on net margins and profits. Although British retailers have faced pressure recently (see Wrigley 1995), they have maintained and expanded exceptionally high profit margins, from 5 per cent in 1980 to 9 per cent in 1990. Between 1985 and 1992, the profit margins of the 'big three' British retailers increased from 5.25 to 8.71 per cent for Sainsbury's, 2.72 to 7.09 per cent for Tesco, and 3.57 to 7.49 per cent for Safeway. These exceptional margins can be, and indeed usually are, interpreted in strict economically rational ways. These realities, however, are outcomes of a contested process of market maintenance and expansion which has to be deeply embedded into the social and political apparatus of the nation state, and increasingly the EU. As with any type of economic activity, and especially in the case of retailing, given its 'near market' location and traditional reliance upon creating markets under highly competitive and inelastic conditions, it has to rely upon and begin to remould the social and political conditions in which it finds itself.

On the face of it, British retailing has been typified by high gross incomes relative to its international competitors through developing efficiencies on sourcing and the rapid development of large stores. This has allowed the efficiencies to be captured 'in house' rather than in passing these directly on to the consumer in the form of lower prices. In addition, looking internationally it can be argued that, for instance, German and North American counterparts have traditionally faced more competition from the discount sector and that British retailers have been more effective at developing distribution technologies.

More significant in terms of these features of the uneven development of international retailing is the speed at which food commodities are rotated. Retailers are not interested in direct backward or forward integration. Indeed they prefer, increasingly, to reallocate the considerable risks in food procurement and quality maintenance to the other actors and agencies involved. Their prime focus is to increase the speed of turnover as well as to demonstrate product innovation and new value added quality. British retailers have given considerable emphasis to reducing the number of 'days stock held', and in balancing off supplier and banking debts by cash cycles which maximise their control of funds and stock. Efficiencies in the 'days stock held' has allowed the time of internal funds rotation to be shortened. In Europe, average days stock tends to vary between 25 and 30 days, but it is significantly lower in Britain. Not only does stock rotation vary internationally, but it also varies intra-sectorally. It increases the efficiency differences between the 'super-league' of retailers and the rest, particularly the ailing independent grocery retailers.

By 1990, five corporate retailers in Britain – Sainsbury's, Tesco, Argyll (Safeway), ASDA and Gateway – controlled 60 per cent of the UK grocery market. By 1997, fewer than ten multiple retailers were responsible for about 70 per cent of retail food sales (Euro PA and Associates 1998). The decade 1986–96 saw the rapid overtaking by the supermarkets of the sale of food goods compared with the independent grocery and butchery sector. Moreover, by 1995 corporate retailer own label products accounted for half of the packaged grocery market shares of the four leading retailers. Food, drink and tobacco sales through the top multiples has continued to expand throughout the mid-1990s (from £36 billion in 1995 to £44 billion in 1997). This has largely been at the expense of the specialist independent sector (that is, the small multiples) and the specialist retailers (butchers, bakers, grocers and dairies). Within a European context, the trend towards rapid concentration in the 1980s and 1990s may be most evident in Britain, but it is in motion elsewhere as well. In France, for example, retail outlets have declined from 200,000 in 1966 to 150,000 in 1990; in 1992, hyper-markets and supermarkets increased their market shares by 20 per cent and 40 per cent respectively, and while they accounted for only 6 per cent of food stores they held a combined market share of 41 per cent. Italy, stereotypi-cally perceived as a very fragmented retailing nation, witnessed a doubling of the number of supermarkets between 1985 and 1992, with the number of hypermarkets increasing sixfold. The number of retail outlets dropped by 15 per cent between 1985 and 1993.

Some key regulatory domains

In the British case the continuing dominance of a small group of retailers, and the absorption of food markets by them at the cost of independent and specialist retailers, is an outcome of social and political factors as much as it is from economic criteria. Moreover, their particular path to success, in rela-tion and sometimes in contrast with counterparts in other parts of the advanced world, has relied upon the particular shaping of the diverse regula-tory conditions that surround them. Regulation in this sense is seen as the process by which power relations (whether within government or beyond) come to be codified and expressed, and through which contested actors and agencies align in order to progress courses of action (see Clark 1992; Hancher and Moran 1989; Christopherson 1993). In the British case, the forms of regulation which have been significant have developed at least at two levels. These can, in summary fashion, be labelled *macro regulation* and *micro regulation*. Both have heavily involved more than simply state agencies, for they have embodied complex networks of public and private actors and agencies. At the macro level the development of the state in the 1980s and 1990s embodied notions of privatised consumption through the privatisation of former state assets (see Saunders and Harris 1990). In this sense, we argue that during the same time period, retailers have at the very least been able to capitalise on the reorganisation of expendable incomes brought forth by the development of privatised consumer culture at the expense of a growing and in consumer terms, disenfranchised underclass (see Marsden and Wrigley 1995).

In short, the development of the neo-liberal state, initially in Britain and the USA but increasingly elsewhere, has provided broad macroeconomic and social conditions which have been conducive to the restructuring of the retail sector along corporate and economies of scale priorities. Underneath this macro economic and political shift has been the uneven development of more micro-national forms of 're-regulation'. These have provided the basis for specific nationally based retail capitals to further shape the food supply and provision systems. Food consumption has not been homogenised by this process. Rather, it has become more reliant upon the differential development of nationally based retailers and a redefined state regulatory role. We have therefore witnessed the development of new spatially uneven regulatory domains which begin to foster and promote the corporate retail sectors and the modes of consumption they construct. So how is it that retailers, particularly in relation to the nation state, have managed to maintain their market power and, moreover, their public legitimacy in the face of potential moral and political crises which have surrounded food consumption?

By the early 1990s in Britain, following the passage of legislation such as the Food Safety Act 1990 and the policy document *The Health of the Nation* White Paper (Department of Health 1992), there was a discernible switch in public policy to shift responsibility for food matters to the retailers; the major retailers have clearly become significant actors in the promotion and implementation of, for example, health policy. As a result, in a very real sense, they act on behalf of the state in delivering consumer rights and choices. Reciprocally, the regulatory state has become critically dependent upon the continued economic dominance of the retailers in their role as the major providers of quality food goods. The state has shifted from its postwar Keynesian position of regulating structures of provision through macro-corporatist arrangements with producers and manufacturers. Instead, more nuanced micro-corporatist relations (see Chapters 6 and 7), particularly with the service and retail sectors, provide opportunities to reorientate and develop food markets and consumption, establishing more *inclusive* rather than *universalistic* provision systems.

The 1990s witnessed a new period of retail-regulatory activity which was broadly designed to maintain markets for the retailers and provide a legitimate mode of consumption for the state acting on the basis of a reconstituted 'public interest'. In the international context, this begins to demonstrate considerable advantages for British retailers over their foreign competitors, and it goes a considerable way in explaining their growing national and international market power. At the more micro level of national regulation, there are at least four regulatory domains in which British retailers have managed to maintain and shape their competitive space and flexibility; we explore these briefly below. It should be stressed that these are very much contested regulatory terrains, and ones which embody different dimensions of public and private interests. In this sense, they are constantly evolving and reliant upon the relative strengths of the main actors and agencies involved.

Competition and pricing policies

The key dynamic here has been the efforts of the discount retailers to establish their supply chains, often in the face of opposition from the corporate sector. For example, in 1994 Aldi complained to the Office of Fair Trading that it was facing discrimination in its purchasing of products from some food manufacturers. The company raised the public profile of the issue to promote and reinforce its low-cost image. Other aspects of the competition and price dynamic include the deregulation of milk prices following the abolition of the Milk Marketing Board, allowing the supermarkets to increase their milk sales by undercutting the price of the pint delivered to the door and sold in the corner shop.

The promotion of the internal market in Europe and the stimulation of competition within and across national borders is creating opportunities for retailers to develop new centralised systems of distribution and supply chain management. For instance, amongst the large retailers it is no longer commonplace for the vast range of manufacturers or wholesalers to deliver directly to the retailers. The norm now is to deliver to a central warehouse, thus reducing the number of deliveries. This rationalisation reduces the need for retail storage space and wastage, and provides the retailer with more control over the flow of foods in the supply chain. These relationships between retailers and their suppliers are, however, increasingly the focus of attention by national and EU competition authorities. The EU has been currently reviewing its vertical restraints policy and it is increasingly recognised, given the process of both concentration and centralisation of corporate retailing, that particular agreements and contracts between suppliers, distributors and retailers can be used to partition the market and exclude new entrants who might intensify competition and lead to downward pressures on prices. It is recognised that:

> agreements between producers and distributors (vertical constraints) can be used pro-competitively to promote market integration and efficient distribution, or anti-competitively to block integration and competition.

The price differences between Member States that are still found provide the incentive for companies to enter new markets as well as to erect barriers against new competition.

(European Commission 1997: 1)

The evolution of the Single European Market and the rise of concentrated and centralised retailing mean that, as far as food supply chains are concerned, there is anything but a free or level market in the provision of foodstuffs. What Europe is witnessing are new forms of restrictive competition based upon supply chain contracting. Here, the competitive relations between firms and between their different supply chains become an important focus of concern. As the Commission has recently argued:

The long term viability of any individual member of a supply chain is becoming increasingly dependent on the ability of the entire chain to compete with the chains of other operators. For this reason, members of a chain may seek to influence its functioning. While many producers lack the financial resources to sell direct to final consumers, wholesalers and/or retailers are used primarily because of their superior efficiency in making goods widely available and accessible to targeted consumers. In the face of competition from large distribution undertakings, many small and medium-sized undertakings have defended their market shares by associating together in a network which grants them access to efficient logistical structures.

(European Commission 1997: 10)

Thus any notion of 'the level playing field' of regional internal markets is confronted with the emergence of uneven retailer-led supply chains which are serviced by particular groups of producers and suppliers increasingly on the basis of the quality and consistency of their products. These supply chains begin to shape competitive space in new ways; quality, choice and credibility become key touchstones in the private and public regulation of these; and the exploitation of retail spending markets (for instance, representing over 53 per cent of total household consumption in the EU) becomes the driving force for supply chains.

Protesting groups of farmers in the UK in December 1997 and growing disquiet at the possibilities of further mergers in the UK retailing sector begin to indicate that the legitimacy of corporate retail power and their supply chains can no longer be taken for granted. The concerns expressed in the recent European Commission Green Paper are also specified in a recent UK Office of Fair Trading Report (OFT 1997). This indicates that vertical restraints are now very important particularly between retailers and suppliers, that these are now retailer-led rather than manufacturer-led, and that the actual form and relationships under vertical restraints have now become much more complex. This latter point means that it becomes increasingly difficult for regulatory authorities to monitor the process. Vertical restraints are now more subtly applied, not just associated with aspects of exclusive supply, refusal to stock or delisting, minimum supply levels or sunk facility requirements on suppliers. Now, the development of more specific agreements and contracts (for instance concerning specific quality parameters) means that the boundary between vertical intervention and vertical integration has become much more blurred.

For regulatory authorities, this means that even if there is the political will to investigate vertical restraints as a key aspect of retail power (so far, despite EU concern, the new British government has been reluctant), then it is going to be very important to adopt a methodology which takes into account *all* the ways in which retailers and manufacturers intervene in each other's behaviour, and not just in the more formal vertical restraints. Regarding any future state intervention in this regard, it is now recognised that there are significant differences in the nature of retail competition and market power in comparison with other sectors, given retailers' pre-eminent position in supply chains, the wide range of dimensions involved in retailer competition, and the peculiarity of the ways in which vertical relationships are expressed (OFT 1997). Moreover, such vertical restraints need to be seen in the context of the two other key competition issues: pricing and merger issues.

Planning and the environment

In many parts of the EU and Japan, there are strict controls over store size and location (see Guy 1998). As a result, the benefits of economies of scale have been reduced and the position of small retailers enhanced. The strictest protection of the smaller retailer occurs in Japan. The *Large-scale Store Law 1974* (see Rabobank Nederland 1994) has regulated the construction and opening of stores larger than 3000 m^2 in Japanese metropolitan areas. The Law's revisions in 1978 lowered this even further to 500 m^2, and there is considerable pressure on government to relax these constraints further. The British planning system has been much more liberal, at least up until the early 1990s. The publication of the Department of the Environment (DoE) Planning Policy Guidance (PPG6) note (a key advisory document for local authority planners and developers) in 1993 on town centres reflected the growing concerns by the public and the independent retailers of the effects of the out-of-town developments. Nevertheless, rather than being seen as a potential threat to the expansive strategies of the major food retailers, the revised PPG6 can be seen as coincident with the thrust of the post-property crisis store development programmes, particularly those of the 'big three' (Wrigley 1998). The changed conditions of competition, and an acceptance of the need to correct the serious overvaluation of the property portfolios of the major firms, has resulted in a significant scaling back in out-of-town developments and a reconsideration of selective in-town or edge-of-town locations.

In addition to uneven planning regulation across nation states, a more recent and growing area of policy concerns waste management and the environment. No retailer can now ignore the environmental concerns of the regulators or of the consumers. One challenge, tackled in rather different ways by retailers and governments in the EU, has been the need to respond to the Directive on packaging waste. Here, the initial efforts of the British government to put the onus of meeting recovery and recycling targets on industry led to considerable difficulties, and the opportunities of self-regulation have lost out to the advantages which firms perceived in passing costs back along their supply chains.

Food law

Packaged foodstuffs in the USA must comply with food labelling, additive, flavouring and packaging legislation. In the EU, the respective pieces of legislation are in various stages of implementation at the member state level, and/or are being modified by the EU Commission. Basic EU product labelling requires information about additives and ingredients, weights and measures, use-by dates and other requirements such as instructions for use and storage if necessary. British retailers, either independently or collectively through the British Retail Consortium and its European partners, are actively attempting to shape this legislation rather than to resist it outright. This is very much a policy of negotiation and accommodation rather than outright resistance. From the perspective of the British retailers and government, the best interests of public and consumers are served in 'regulating last', and by maintaining corporate flexibility. The 'regulating last' principle assumes that regulation is a last resort, after all other possible options have been explored and the full 'costs of compliance' have been accounted. Moreover, in terms of any transnational expansion plans, retailers need to be able to promote this *modus operandi* into the logics of other nation states (such as those in Eastern Europe) through the promotion of harmonisation and de-regulation.

Food quality

In terms of food quality, it is clear that baseline legislation from either national governments or the EU provides only part of the picture concerning the delivery of quality food goods. The EU Directive on the hygiene of food-stuffs (Council Directive 93/43/EEC) puts greater emphasis than before on the policing of food quality on private businesses. As we shall see in Chapter 6, the definitions and conflicts between private and public interests are most explicit in this area of regulation, both at the professional level (i.e. between environmental health officers, public analysts and private retail quality specialists), and between the different arms of the two main government ministries involved (i.e. Department of Health and MAFF). Food safety and nutrition form a significant brief for the Department of Health. Following the publication of *The Health of the Nation* White Paper in 1992, several task groups were formed involving the food industry, government and scientists (and some representatives from consumer groups) to further refine sets of quality guidelines with respect to more healthy diets. For instance, the Nutrition Task Force set targets for the reduction of fatty acid consumption and obesity, and attempted to 'persuade' manufacturers and retailers to undertake 'fat audits' as part of their quality control procedures. Documents emanating from these public-inspired initiatives (see Department of Health 1994a) caused considerable concern to the retailers and manufacturers, who are constantly in the process of creating and designing new products but who wish to do so on their own terms as much as possible. In particular, the 'National Food Guide' published by the Department of Health in 1994 suggesting the need to modify diets towards the consumption of fruits and vegetables, bread and breakfast cereals, potatoes, rice and pasta, created a considerable storm. In addition, the scientific evidence emerging from the Cardiovascular Review Group (Department of Health 1994b) exposing the problems of fatty diets, obesity and considerable socioeconomic and regional disparities in food consumption and health, has led to heightened awareness of the significance of food quality and food choice, and to questioning what the role of government *vis-à-vis* the suppliers of foods should be. In a somewhat telling defence of the Department of Health's new initiatives, the Parliamentary Under-Secretary of State argued:

> No government should be a disinterested observer of an unhealthy nation. It is rightly a matter of concern, for example, that Britain has among the highest incidence of coronary heart disease in the western world. Avoidable ill health is not just a drain on the quality of people's lives, it imposes an enormous cost on both the NHS and the economy as a whole. For these reasons the Government is committed to meeting the targets set out in the Health of the Nation White Paper.
>
> (Press Release, August 1994)

Shaping regulatory domains: the evolution of competitive space

These new guidelines and recommendations are beginning to frame new regulatory parameters for the food industry and retailers in particular, even though it is unclear how they will be implemented. As we will analyse below, food retailers are having to engage and promote their powers of legitimacy both inside and beyond government. Probably more so than in the other regulatory domains outlined here, food quality regulation most clearly exposes retailers to the very risks and responsibilities their particular dominant location in the food supply chain bestows upon them. Moreover, in conditions where consumers are more conscious and sensitive about the food they eat, this area of regulation holds important competitive implications for the retailers.

Further, the increasing role of retailers as mediators between the consumer, the state and food manufacturers makes them a critical agent in the delivery and modification of consumption patterns. Much of the state apparatus recognises this. In conditions where there is a growing consciousness about the quality and origins of food goods, both the retailers and the state have to constantly redefine their relationships with each other. When public concern mounts, a major question for government becomes how to influence a corporate sector, which, by its definition, owes some of its rapid empowerment to the deregulatory initiatives of the state; while for retailers, there remains the question of how they can create and maintain competitive space. The role of government is considered further in Chapter 6; below we explore in greater depth the meaning of 'retail competitive space'.

Our starting point is the recognition of the need to go beyond the rather dichotomous assumptions in the literature concerning the role of retailers as private operators in the market place *per se* and the role of the state and regulation as a separate entity. While their origins and objectives often stem from quite different quarters, we have to recognise – taking what Clark (1992) calls a 'real regulation' stance – that it is how both come together in particular arenas which is analytically important in understanding retail development. Thus, we need to focus on how regulatory change and contestation and new competitive conditions occur out of the *hybridisation* of market and state power. In this sense, the notion of 'competitive space' indicates the dynamic outputs of this hybridised regulatory activity. Second, once regulation is placed as a central and active context in which the space for private capital and the rules of exchange are forged, it is possible to appreciate that this can only lead to a dynamic process of uneven development. The survival of one or more retail firms usually means the decline of others. Market rules, continually developed and shaped, hold distinct costs and benefits for retailers and consumers. Hence, it is not just the innovative economically rational activity 'of the firm' which generates its potential dominance in the market place; rather, it is the way the firm interacts and interfaces with other firms, consumers and regulatory bodies over time and space, forging different regulatory alliances which begin to recreate the legitimate conditions for the pursuance of its increasingly diverse activities.

The reconstitution of regulation as a product of both 'state' and 'market' relations and the reality of uneven development as an outcome of these relations become, therefore, important reasons for pursuing the concept of 'the making of competitive space' (see figure 3.1). Many of these are expressed in the emerging literature concerning the 'new' retail geography. There are at least three key dimensions (*spatial*, *intra-sectoral* and *supply-based*). It is important to distinguish these dimensions because of the differing (and potentially somewhat chaotic) meanings which can be attributed to the

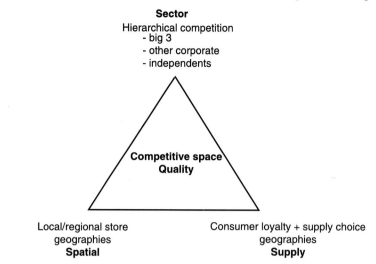

Sector
Hierarchical competition
- big 3
- other corporate
- independents

Competitive space
Quality

Local/regional store   Consumer loyalty + supply choice
geographies   geographies
Spatial   **Supply**

Figure 3.1 Making competitive space

notion of 'space'. In addition, our evidence, which we report here and also in Chapter 6 and Part III, suggests that in each of these dimensions, the variable definitions and constructions of food quality become increasingly important.

Probably most popularly documented in the policy and scholarly literature are the actual spatial battles associated with the competitive siting of retail and warehouse outlets (the so-called 'store wars'). This is very much played out in relation to the changing nature of planning legislation and guidance, and has recently involved retailers in collaborating under the auspices of the Retail Planning Forum. This essentially local process of gaining and holding on to the literal end of competitive space provides a major dynamic in the provision of retail stores (Langston *et al.* 1995), and the restructuring of consumption spaces more generally. There is now an expansive literature on the siting of retail stores, partly stimulated by the role of applied geographical analysts in assisting retailers in developing ever more sensitive appreciations of the social geography and consumer profiles of particular towns and cities. The corporate retailers have traditionally targeted the higher social class areas, with the discounters following up with locations adjoining them or targeted at working-class areas. This competitive process has served to solidify the socio-spatial polarisations to such a degree that some commentators have begun to identify 'food deserts'. These are areas bereft of superstore development and ailing independent grocers forced to charge high prices to their low-income customers (see Institute of Public Policy Research 1994; National Food Alliance 1995; Dobson *et al.* 1994). The playing out of the 'store wars' battles, presided over by an often compliant planning system which has attempted to negotiate planning gain developments on their rights of acceptance of development, is serving to heighten the spatial disparities in food provision, leading to uneven and large consumption spaces with their own geographies of plenty and scarcity.

However, we should recognise that to focus upon the actual new siting of stores and their effects for consumption is only one side of this spatial dimension in the remaking of retailers' competitive spaces. Of increasing significance are the 'loyalty geographies' of consumers. As the Institute of Grocery Distribution (IGD) (1996) has argued from extensive consumer research, it is the issue of consumer loyalty and confidence which now becomes a major focus for retailers. Major innovations concerning the development of SMART cards, membership deals, local event sponsorship, free transport and the possibilities of home deliveries and electronic communication are being tried and tested by the corporate retailers. These strategies represent a coming to terms with their ongoing need to regenerate their catchment loyalties in the spaces in which they have managed to situate themselves. Here, the supply and quality of the retailers' delivery becomes of critical importance in maintaining their competitive space *vis-à-vis* their local competitors. This is illus- trated by the following quotations from interviews with senior retail staff:

> The two concepts that we try to focus upon in regard to the customer in our business: one is the sense of belonging – we want to create a sense of belonging amongst our customers. The second is – and nobody is doing it well yet – is creating a sense of excitement – the thing is, shopping is largely a boring occupation, and anything you can actually do to liven it up ...

> I mean we like to think that we can get quite close to our customers. We are in the business of trying to communicate really clearly to them. Because we are in a market where there is abundant competition, we sure as hell do our best to do it well because we desperately want to acquire and maintain loyal customers. We are not trying to rip people off. Our job is to find customers in a saturated market and sell them goods equivalent to the price of a small family car every two years. You have got to develop that level of loyalty.

If the 1980s and early 1990s witnessed the growing physical and social presence of the corporate store into local urban spaces, the more recent period, while continuing this trend (Langston *et al.* 1995; Guy 1995), is supplemented with a growing concern with harnessing catchments of consumers around particular value, quality and loyalty criteria. Redeeming the sunk costs of site expansion increasingly engages the retailers in the local duty to 'serve the consumer', attempting to project an image and a reality of local customer service above and beyond their local competitors.

If, then, the local maintenance of competitive space as literally a spatial dimension becomes a major arena for retail corporate strategy and uneven development, so does that concerning supply. Power *along* supply chains and the increasing dominance of retailers in negotiating quality control in contracts and agreements, becomes a major feature in maintaining competitive space, between themselves and with food producers and processors (see Hughes 1996; Doel 1996). The role of supply chains in regulation is developed further in Chapter 6.

The British system of supply chain management is generally viewed as more 'retailer led' (Doel 1996; Hughes 1996) than that of the USA, for example. The growth of retailer own-label, reaching 50 per cent of the 'packaged' grocery market in the UK by 1995, and the weaker market position of manufacturers compared to their US counterparts means that:

> The composite character of the manufacturer interface is one of dynamic complexity ... There is a widespread consensus that those own-label supply networks created by proactive retailers and characterised by intensive interaction are assuming progressively greater quantitative and strategic significance.
>
> (Doel 1996: 62)

In the British case, the evidence suggests that these relationships not only begin to empower the corporate retailers over their upstream suppliers (that is, constructing a supply-led form of competitive space), but that the particular supply relationships developed provide a basis for creating intra-sectoral competitive spaces between the Big Three retailers, the other corporates, the discounters and the independent sector. This is an argument that we examine in more detail in Chapters 6 and 7 where, as we shall see, this is particularly associated with food quality concerns in a period when the attribution of food quality standards and regulation becomes much more sensitised, both by retailers intent on enhancing their loyalty rating, and by 'careful' consumers and agencies concerned with the increasing risks and responsibilities of food supply and provision. The development of *quality* supply management thus becomes highly innovative, with exclusive supplier relationships developed between producers, processors and retailers.

There is, therefore, considerable evidence developing of the multi-dimensional nature of the making of corporate retailers' competitive space, along at least three interrelated dimensions: *intra-sectoral*, *spatial* and *supply-based*. Retailers' strategies have to deal with these simultaneously over time and space. These parameters begin to set a more sensitive analytical context for studying the uneven development of retailing, and particularly how it is maintained under intensely competitive conditions. In this sense, this begins to analytically frame the social and political bases of the new competitive conditions in which retailers find themselves.

Sustaining retail power

Retailers thus have to engage in multidimensional strategies in order to maintain and sustain their competitive space. These involve pursuing individual strategies with government and the EU, and joint arrangements with such bodies as the British Retail Consortium and the all-party group of MPs in Westminster. At the European level, their participation is growing through EuroCommerce. These sets of relationships, following a broad private-interest regulation model which we develop in Chapter 6, could not occur, however, without significant changes in the state's handling of retailer and food concerns. The increasing significance of the corporate retail success story for deregulatory governments of the 1980s and 1990s has been combined in the British context with the gradual decline of agricultural corporatism (Flynn *et al.* 1994) and the independent grocery and butchery sectors. Even more significant has been the increasing realisation by government that consumer concerns over food safety have stimulated a rise of consciousness about the overall management of food supply. Retailers, given their pivotal position in supplying choices, and the enhanced degrees of freedom conferred on them by government, become acutely important for the legitimation of the state and, more specifically, for the management of the food system on behalf of the state and the consumer interest. With millions of consumers passing through their doors each day, and with the myriad of products purchased by them, retailers can project a more critical custodial role to government and the consumers, and in so doing help to marginalise other forms of consumer representation, as we shall see in Chapter 8. In addition, in a period when state regulation has to focus more on the consumption process, as opposed to production, the intensity of retailer–government relationships takes a new turn.

In government circles, the main ways these relationships with retailers are expressed are through interactions with MAFF and the Department of Health. From the retailers' position, and indeed for many of the officials, MAFF's role (particularly under the former Conservative government) was increasingly seen as a sponsorship role for the food industry in general, and retailers in particular. An executive for the British Retail Consortium put it this way:

> Oh yes, in the time that I have been dealing with them, that is over eight years, I think they have definitely become more open – particularly MAFF. Actually I think the food industry has always had a good relationship with them, because MAFF needs such a relationship. You see MAFF is sponsoring the food industry. I mean that is probably not a very good word. MAFF is supposed to look after the food industry's interests, and to give help and to iron out problems in areas where the government can.

While one civil servant claimed:

> It's very much government policy these days that you must try to avoid regulating business. I mean the point is that actually present ministers neither believe in regulation, nor do they believe in spending money. So in a way, they inhabit the middle ground where they believe in encouragement, facilitation, knowing a lot about the industry, rewarding good practice by drawing attention to it and so on. That's the sponsorship role in a sense. It's slightly new in a way. It's definitely new from when I was a young civil servant when we were emerging from a period of very strong regulation.

A highly significant outcome of retailer–government interactions is that the former are increasingly regarded by the latter as much more viable and legitimate representatives of the consumer interest than consumer groups. Indeed, retailers are in some cases seen to moderate government policy on behalf of their consumers. For instance, it is argued that while the 'scientific evidence' supporting food irradiation from MAFF is sufficient to grant application of the technique on a range of products, this has been restricted by the retailers because of their recognition of consumer consciousness in this area. Retailers can moderate as well as innovate government policy, and MAFF gives them a special ministry, over and above other industries, in which to influence and manage their competitive spaces and constructions of consumer interests.

The individualist and voluntarist system of policy mediation pervades retailer–government relations, and it is also clearly expressed in the Office of Science and Technology's Retailing and Distribution, and Food and Drink foresight reports (see Cabinet Office 1995; Office of Science and Technology 1995). The self-regulation of consumers is juxtaposed with voluntarist, private-interest regulation on the part of the retailers. Government's role is to legitimate and codify these systems, demonstrating their value in sponsoring a competitive economy. This also, not simply by default, reinforces the competitive spaces between retailers discussed above and elaborated further in Chapter 7. It also allows, for instance, some of the 'Big Five' to communicate low-fat 'healthy' food lines, providing marginal gains in the competitive spaces between retailers on the basis of food quality criteria. In this sense, regulatory policy provides the competitive wherewithal for some retailers to gain market share over others. This tends to reduce the status of universalistic principles of public health provision. As one health official said:

> But we are not talking here about certain people at risk. We are talking about moving the whole population. I mean the entire population need to reduce its fat intake by about 35 per cent ... The whole point of the age of population approach is that if you compare the UK curve with

the Japanese curve, you just need to shift the whole lot down, and everyone will benefit, and you can also target the high risk groups at the same time. But it is a philosophy that a lot of people have problems with. There is evidence epidemiologically that you should shift the whole population, and that's what we are trying to do. There are also things we are trying to do through the MAFF groups, like changing the composition of the products on the shelves, not a lot, but just a bit. We don't need claims that such and such a product has a 25 per cent reduction in fat, but that we need a small reduction of fat *in everything*. So that everyone can go out and choose a low-fat diet much more easily. I mean I think there is just a difference in philosophy between the population approach and the targeted one. We can't do both.

Conclusions

Our analysis in this chapter has demonstrated how corporate retail power is evolving in the context of particular socio-political conditions. Corporate retailing in Britain, and we expect elsewhere, is a highly embedded activity, not only in the consumption process itself, but also in the socio-political cultures of nation states. In addition, retailing is not simply reactive to state activity; as the evidence shows, the current nature of state relationships encourages diverse forms of participation in the making and implementation of policy. This in turn, provides a basis within which new rounds of uneven retailer development can take hold. Retailers and the state, projecting their representations of the consumer, become powerful actors in influencing consumption and its regulation. Participation in the regulatory domains outlined earlier provide major opportunities for retailers to shape their own and their competitors' market spaces. In social and political conditions which continue to emphasise choices, quality and consumption as elements of citizenship, such participation is likely to become all the more significant in maintaining corporate retail power. The links between citizenship and food rights are explored in the following chapter.

While the scholarly literature has begun to give considerably more emphasis to the process of 'real' regulation and influence on the economic landscape (see Clark 1992; Flynn and Marsden 1995), the analysis here demonstrates that much more work needs to be done in understanding how actors and agencies participate in the process on a continuous basis. We seek to fill at least part of this gap in the remainder of this book. Regulatory domains, such as those associated with food quality, have to be filled with contested knowledges and arguments which have to gain levels of relative legitimacy and authority in order for them to become powerful, and therefore able to influence policy, markets and corporate governance. As we shall see in Chapter 6, within the operation of the British private-interest regulation model, particular conceptions of the consumer interest are articulated inside the regulatory domains. The outcomes of these processes of contestation, for the time being at least, seem to continually empower the concentration of the British corporate retail sector not only as a vanguard economic sector in its own right, but as the late modern custodian of public food supply in a period of considerable questioning about the type of quality of food we consume. It is to the emerging issue of consumption rights that we now turn.

4   Citizenship, consumption and food rights

Introduction

In the previous chapter, we addressed how British retail capital sustains itself through its embeddedness in socio-political structures. The engagement of retailers in regulatory activities is pursued further in Chapter 6. In this chapter, we explore the ways in which the rise to prominence of retail capital is associated with a reformation of rights to food. For the maintenance of their own competitive position *vis-à-vis* other retailers and others in the food system, corporate retailers engage in the constant construction of the individualised 'consumer'. In order to critically explore changing rights to food we draw more heavily upon the literature of citizenship than of consumption. Citizenship provides a useful starting point for our work because of its more explicit links to rights and access to services than the more fragmented literature on consumption. Gabriel and Lang (1995) also draw attention to the links between consumption, citizenship and rights, but while pointing out the tensions between notions of the citizen and the consumer – the former a political concept and the latter an economic one (Gabriel and Lang 1995: 175) – they do not pursue the links between these two concepts and that of rights.

As we shall see, rights in relation to food have undergone significant changes in the postwar period, and with this change has gone an increasing differentiation of access to quality food. The more 'political' terms of citizenship and rights are more easily linked to those of regulation. Both the literature on citizenship and consumption are, however, somewhat deficient in their analysis of regulation which is at best left implicit and at worst ignored. Yet as we show, it is the changing nature of food policy and regulation which we explored in Chapter 2 that plays such a key part in configuring rights and thus our notion of citizenship (in relation to food) and consumption practices. Similarly, both literatures tend to be rather weak in their analysis of the state and especially of the way in which it may be restructuring itself (though the work of Peter Saunders is a significant exception here). Again, though, an understanding of the role of the state and the way in which it may be changing is of central importance to the way in which food rights are reconfigured and by whom. More specifically, in this chapter we outline the ways in which changing notions of food rights are linked to those of a more privatised consumption perspective; this provides a considerable contrast to the early postwar period, when state conceptions of food rights were dominant.

Citizenship and food

The classic work on citizenship in a British context is that by Marshall (1963; 1965; 1981). Marshall was primarily concerned with the growth of citizenship as expressed in three dimensions: the civil, the political and the social. The development of rights in each of these three spheres was linked to a particular historical period. The eighteenth century witnessed a significant development of individual civil rights; the nineteenth century saw the growth of political (i.e. electoral) rights; and the twentieth century has witnessed the spread of social rights (i.e. claims to welfare) (Turner 1990: 191). In his enquiry into citizenship, Turner explores in detail the validity of key criticisms that have been aimed at Marshall's work. Many of these, he claims, are misguided or based on a misunderstanding of Marshall's ideas. Nevertheless, Turner (1990: 193) does point out that within Marshall's framework it is the state which provides the key means for the development and maintenance of social rights, but that his analysis of the state was weak. For our understanding of food policy and food rights, and their implications for the consumer, the state plays a central role; but so too does the unfolding relationship between the state and key elements in the private sector, notably food retailers. As part of the restructuring of the contemporary British state, at least in the area of food regulation, it is involved in sharing its authority with major retailers, and this we explore in greater depth at both a conceptual and empirical level in Chapters 6, 7 and 9. As we shall see below, this has significant implications for our understanding of the formation of citizenship.

Turner points out that not only is Marshall's concept of the state inadequate it is also outmoded (Turner 1990: 195). In order to update the relationship between state and citizen and to develop a comparative framework, Turner draws upon the work of Mann (1987). Mann is able to present a much more sophisticated analysis of the process of citizenship formation and the way in which it may vary between different nations by conceiving of citizenship as an element of class relations. Mann, however, according to Turner (1990: 197–9), despite presenting a more dynamic analysis of the formation of citizenship is still partial in his approach. Most significantly Turner (1990: 199) argues that

> Mann can only conceive of citizenship being handed down from above (for example by the state) such that rights are passive. Thus, citizenship is a strategy [by the state] which brings about some degree of amelioration

of social conflict and which is therefore a major contribution to social integration.

As Turner points out, though, such a top-down perspective is probably historically inaccurate and socially exclusive. It 'precludes, or restricts, any analysis of citizenship from below or as a consequence of social struggles over resources' (Turner 1990: 199) and so ignores, for example, the environmental movement. Moreover, as we shall see in Part III, our work suggests that there is at least one possible further source of 'right' formation. That arises from the interactions between state and private capital and occurs because of the restructuring of the state and constellation of retail capital. To portray the reformulation of food rights in the postwar period as simply a ruling class strategy would be naïve, as it would gloss over the tensions and contestations between retailers and government and between retailers and other sections of the food system as we have shown in Chapter 3. Similarly, to portray such rights as the outcome of radical struggle would be to grossly overestimate the significance of the food consumer movement (see Chapter 5).

Turner's development of the notion of citizenship builds upon the above/below distinction and adds to it a public/private dimension (Turner 1990: 200). The outcome is a typology which seeks to place citizenship into one of four categories: revolutionary, liberal pluralist, passive democratic and authoritarian democratic. Turner is keen to introduce these categories 'to emphasise the argument that citizenship does not have a unitary character' (Turner 1990: 201). While we would agree with these sentiments we would also like to develop and apply this line of thinking rather further than Turner appears to. He seems to believe that the typology is a means of identifying national characteristics of citizenship but it is equally plausible that such heterogeneity may be found within nations during different time periods and around different policy spheres. For example, the state and social movements constitute themselves in different forms and articulate and agree on rights demands. As we show below, changes in food policy specifically have been inextricably linked to a significant shift in food rights in the postwar period.

Citizenship rights are, therefore, about forms of entitlement and carry with them expectations. Citizenship thus carries with it a sense of 'material circumstances experienced by individuals, households or groups, and to sets of ideas and claims about rights or obligations. These two dimensions overlap, and are hard to separate' (Harrison 1991: 209). The ways in which those rights are constructed (i.e. the latter of these two features) are our principal concern and are outlined below.

But as Turner (1990: 211–12) points out, the concept of citizenship is a dynamic one. He identifies tensions between globalism and localism as reconstructing notions of citizenship, but equally as important may be the tension between productive and consumptive processes. It is as well to remember that it is the state which plays a key role in the construction and maintenance of rights and that it is heavily involved in the regulation of both production *and* consumption practices. Indeed, in his commentary on Turner's paper, Harrison (1991: 211) argues that within the consumption arena citizens are likely to experience quite different levels of treatment. As he puts it, 'citizenship in terms of daily life is likely to be a quite varied experience involving differential incorporation in industrial and consumption terms' (Harrison 1991: 212). What we are able to show in relation to food is how such differentiation has developed through time by linking together producer, consumer and state interests in the construction of rights. In essence, we build upon our argument in the preceding chapter to show that the rise of consumption practices has tended to undermine the traditional role of the state and led to the construction of more privatised (and differentiated) forms of rights to food provision.

Food consumption

As Warde (1990) has pointed out, much of the sociological literature on consumption is in fact the study of consumer*ism*. The formation of tastes, the link between status and purchasing, and between purchasing and personal gratification have all been well covered and have been reviewed by, for example, Warde (1997) and Gabriel and Lang (1995). It has led to a burgeoning and diverse series of works on consumer culture (see also Lury 1996). Rather than focus on the food consumption experiences of individuals or even groups of consumers, our interest here is in *rights* to consumption: in other words, those entitlements that help to structure the choices that consumers can make. This suggests a rather different approach in which analysis shifts to the ways in which choices around food supply are constructed, and the centre of analysis becomes the creation of food choices for consumers. This shifts attention to the activities of retailers and food manufacturers, consumer groups and the state. Obviously the state plays a key role in the food supply domain because it establishes what is and is not permitted in terms of hygiene, content and production and so on. The restructuring of the state, however, casts traditional responsibilities into a new light and leads to the casting off or reformulation of others. As such, at least within the area of food, it leads to the striking of a new balance of service provision between public and private sectors.

One of the few commentators to have tackled the issue of the changing nature of consumption, service provision and rights is Peter Saunders. His early work on the sociology of consumption (1986) has been followed by analyses of the political and social implications of the privatisation of services (Saunders and Harris 1990). This shows how the privatisation of formerly state-supplied goods and services can have highly variable consequences for consumers. At an abstract level they present a typology to distinguish between two key dimensions to privatisation: whether it is private interests or consumers who take over responsibility for goods and services when the state withdraws; and the nature of the change, whether it involves the selling of assets or the relinquishing of control of powers. Together, these two dimensions create a fourfold typology as illustrated in Table 4.1.

The work of Saunders and Harris clearly shows what had been anticipated in the discussion on citizenship, namely that there is likely to be increasing differentiation of consumption rights and that, given the nature of state restructuring in the 1980s, consumption is likely to be a relatively more significant process than in the past. Saunders and Harris are interested in the changing provision of goods and services and make no mention of regulation. Regulatory practices can change, though, as a result of privatisation; the water industry in Britain is good example here. More subtly, as we argue, changing regulatory relationships are taking place with regard to food as an outcome of the restructuring of capital and the state. Saunders and Harris's model brings out well the political and sociological implications of such regulatory change. The growing private-interest regulation of food is linked to liberalisation (cell 3) and the differential ability of consumers to purchase different quality foods to marketisation (cell 4). To apply Saunders and Harris's (1990: 60–1) argument to food regulation, liberalisation refers to the state retaining an ultimate responsibility for safety and standards but sharing the organisation of regulation with private (corporate retail) interests. Private-interest regulation is explored in greater detail in chapters 6 and 7. Marketisation, meanwhile, is where the state transfers its expenditure on services to enabling consumers to purchase what they want in the market: 'the state is moving away from providing a service in favour of enabling consumers to buy it for themselves' (Saunders and Harris 1990: 61).

In terms of food 'marketisation', such processes can be increasingly observed in the purchasing actions of consumers, where decisions about food safety and quality are ever more closely tied to the retailer and the assumed ability of the consumer to choose. As the ability of retailers to regulate their supply chains is variable, so too is their ability and willingness to develop food standards above the state baseline. The result is, that based upon consumers ability to choose where they shop, they enjoy differentiated rights to food quality. It is to the way in which these food consumption rights – an element of citizenship – are constructed that we now turn.

Table 4.1 A typology of privatisation

	Change in government's role	New locus of responsibility
	PRODUCERS	CONSUMERS
CHANGE OF OWNERSHIP	(1) Denationalisation	(2) Commodification
CHANGE OF CONTROL	(3) Liberalisation	(4) Marketisation

Source: Saunders and Harris 1990: 59

Constructing food consumption rights

Perhaps rather surprisingly, there has been little if any linking together of rights to food and citizenship. In part, this seems to have been due to limited analysis of food policy, and in part to an uncritical acceptance of food policy within welfare policy. Indeed, for part of the postwar period such a position was quite understandable. As Chapter 2 argues, postwar food policy was an adjunct of agricultural policy and subject to little public debate or discussion. The role of government was equally clear: it was to ensure freedom from want through the provision of affordable and safe food. There was a sense of a collective consumer interest, whose choice was largely determined by government. Now it is the corporate retailers who play a central role in promoting to individualised consumers their vision of quality and diversity of consumption. This is one of a variety and a hierarchy of standards. Different retailers seek to imbue their products with notions of quality, and to do this they must be able to exercise considerable influence over the supply of products. This is what we term *private-interest regulation* (Harrison *et al*. 1997), because retailers are creating their own standards and operating food safety systems which go beyond that required by more traditional forms of public regulation (see Chapter 7 for the operation of this system). The concept is based upon the assumption that the informed individual is best placed to make decisions on consumption. In other words, the consumer is being given the right, the *freedom to* consume, and this itself entails a rather different notion of citizenship. To understand how such rights can be reformulated, it is necessary to explore the roles of the retailers and other key actors, notably consumer groups and government.

Citizenship rights are the legitimation of social and political claims made on the part of the people. The ability to demand or reject, to provide or accept rights is an inherently political one. Rights themselves are not just important for their entitlements but also for insights that they provide into the political standing of those who possess them (see Jones 1994: 4). One way of trying to understand the nature and dynamics of food rights is to conceive of them as social relations. As Daniel Bromley (1991) has argued in relation to property rights, it is necessary to distinguish between rights holders (in this case consumers) and duty bearers (here farmers, food manufacturers and retailers) in relation to food hygiene, safety and standards. It has traditionally been the role of the state to protect the benefits that consumers derive from these rights. Importantly, though, the state may perform its role in different ways, for example, as a result of regulatory reform. Since it is a social relationship that defines consumption rights, those rights may be subject to contestation. Thus the regulation of food in both its public and private forms becomes a central object of analysis in any understanding of food rights.

As we have argued in Chapter 2, the food regulatory framework is a dynamic one and plays a key part in structuring the rights that consumers hold at any one time. The key constituent elements within the framework are economic interests, although government remains the ultimate arbiter of rules. Government's role is not static and, increasingly in conjunction with private interests, is modifying consumer rights, particularly around contestations concerning quality parameters. This restructuring is determined at the national level but is mainly played out at the local level in the food enforcement practices of EHOs and TSOs (see Part III).

Ministerial thinking on the market and regulation has resulted in a reconstruction of the Ministry of Agriculture, Fisheries and Food's (MAFF) traditional public interest form of regulation which prevailed until at least the late 1970s, and in deepening its relationship with the multiple food retailers. As one official remarked in the mid-1990s:

> The government's position is that the market is, broadly speaking, the best determinant of what happens in industry and business. Ministers see us at MAFF as a group of people who are able, through *releasing powers of regulation*, as being able to facilitate the success of business. The attempt has been made to reduce all regulations to an absolute minimum ... The point is that present ministers neither believe in regulation nor in spending money. [But] they [do] believe in encouragement, facilitation, knowing a lot about business ... that's the sponsorship role, in a sense.

Releasing powers of regulation has not meant that the traditional public sector role of protecting the consumer from health risks associated with food have been subverted. Rather, what has happened is that different sets of rights, intimately linked to private sector provision, have been fostered. Consumers are empowered, are free to make the choices as they see fit. These are competing notions of rights, and are associated with different regulatory arrangements. The public or the private sector may take the lead as appropriate, but in practice they are to be found alongside one another. The resulting tensions within the regulatory framework have been particularly acute at the local level of implementation (see Part III). What we are witnessing, therefore, is the government essentially trying to act as backstop, to ensure basic standards of food safety. Over and above this, the multiple retail outlets are creating additional rights based on their different guarantees of food quality which are available to their customers. This requires that these firms police and regulate their own food supply chains.

For the major food retailers, quality is linked to competition amongst themselves. As a leading figure in one of the major food retailers put it: 'For our customers, we're in the business of offering quality'. He continued:

> Well, we're all competing for the same share of the purse. There is only a finite number of calories that people can eat in a day. Our job is to make sure that it is our calorie that they are buying and not someone

else's. So, we only make our profits by satisfying our customers, and we have to discover what it is that satisfies them ... They're saying that there is a particular level of quality that they want, and we have researched this exhaustively, and every product that we produce goes through customer research to find out whether it is of the right quality.

To be sure of the quality of a product means that the company has confidence in the ability of food manufacturers to deliver appropriate standards. The interviewee then outlines the procedures involved:

The work that we do with the manufacturers – we do not make any products ourselves, but prior to anybody making any food for us, they have to satisfy one of our food technologists that they have complied with the criteria that they have set out in their quality management system manual which involves an audit ... We devise this, in consultation with the suppliers who are involved. Sometimes suppliers say, well that is very interesting but we can't do this and we can't do that, so this consultation process is important. There is a lot of prior consultation. At the end of the day, there is a set of criteria, a set of standards, that we have set out for any manufacturer, and if they can't adhere to those standards in the audit, then they are not in the frame. This has to be sorted out before we start on the negotiations involved in buying stuff. Any buyer in this building knows that he/she does not go to any factory unless they have been formally approved. We will definitely not accept any food from unapproved sources anywhere by anybody.

So once the supplier is approved we start to talk about the product that is to be developed. In that development process, there will be customer research ... through our stores. We have market research agents in our stores asking customers to taste food in kitchens blind, to satisfy quality criteria, so that when we have a satisfactory customer report then we can market the product. So the technologists are really the key people who work in partnership with the producers to design, develop and procure the products. So the quality management process is governed by technologists effectively. They are backed up by laboratories here. We have got consumer kitchens here. We have got sensory panels, fragrance panels, wearer trials of clothing, packaging laboratories – and the whole thing that backs it up by saying there is a need for subjective information to meet our objectives. But what the analysts do is to check the competence of our suppliers. This is not endpoint testing. We are relying on the laboratories, to give the factories the information they need.

The competitive nature of food retailing means that products supplied by manufacturers cannot be stored to wait for tests to be made on them, because this would be an inefficiency in the supply chain. Instead, as a senior retailer explained, what they will say to the manufactures is that:

> our truck is at the end of your production line waiting for that food. It will be in our depots tonight, our stores tomorrow, and sold the day after. So you had better start sorting out your quality management because what I want from you is confidence that what comes off your production line is OK without endpoint testing. There is no time in the world to interrupt the supply chain with endpoint testing. There is no time; you've got to get the job done on the production line. So we went round and said that we wanted the manufacturers to adopt hazard analysis, critical control point techniques, to apply those principles to food processing, so that what comes off the production line does not need endpoint testing to say it is OK to leave the factory.

The food retailers' ability to regulate the flow and quality of food is thus quite different from that of government which, as we shall see, relies heavily upon local government officials. Nevertheless, the supermarkets' regulatory activities in this sphere, what we have termed private-interest regulation, does derive some of its legitimacy from the government's own regulatory approach. Thus, the Food Safety Act 1990 and the requirement of 'due diligence' put the onus upon food retail outlets to be able to show in a court of law, should there be a problem with food, that quality and safety had been managed. For the major retailers, this simply legitimated what they were already carrying out as good practice. There is also a difference in emphasis between private- and public-interest regulation which we can observe. The public interest is largely concerned with baseline measures. For the private sector, there is an additional element in which quality too has a value from which profit and competitive advantage can be extracted.

The results of the supermarkets' strategies are significant for the choices that consumers make. As a retailing interviewee put it:

> We don't have any control over brands. What Mr Mars does, or what Mr Kellogg does, is up to them. But 50 per cent of food in Britain is bought under own label, and we have total control over our own label – in terms of source manufacture, specification, composition, nutrition, packaging right through to the whole thing. It's totally under our control.

Other major food retailers adopt similar procedures to ensure the quality of the food they sell. But as we illustrate below, the different tiers of food retailing have slightly different ideas of quality and the choices they can offer their customers, and thus consumers experience differentiating rights to food quality. As a very senior figure in one of the second tier of national retailers argued:

Every single supermarket chain will establish its own [quality] benchmark, and that will vary. You will get a different benchmark talking to say, Kwiksave than you would talking to us. I mean our benchmark is saying that, on our own label, we have to have a product that is among the top three in its field. In terms of product development, we tend to be a follower rather than a leader, given the nature of our business. We don't, for instance, go out and develop new areas of eating, we are happy to let Marks and Sparks and Sainsbury's go and do that for us. When they develop something that is a winner, then we will piggy back on it. And, therefore, where there are existing products in a particular market, we will continue to make our own. We will say that we want to be at least as good as the best three. So, our buyers will get samples, and food developers will test them, and they will dissect them and they will create a specimen. Sometimes, we will go to the same supplier, and what we aim to do is to get the product that effectively eats and tastes like the best of them. The message that we give to our suppliers is that we are not interested in you adding cost into a product unless it delivers the eating, taste, and utility criteria.

The point that this retailer was keen to make was that price and quality go together:

You can't persuade the consumer to eat cheap rubbish. But we will seek to make a price point. So ... take ready meals, which is a classic example, we originally developed products which were selling at £1.40 or £1.50 for a lasagne, and this didn't sell. So we now make to a 99p price point, a slightly smaller meal, and slightly less packaging, but of perfectly good eating quality. These sell at an enormous rate. So ... you've always got a price point, as opposed to price, in mind.

In short, quality is constructed by retailers and manufacturers. As such, there will be variations in the quality of products between different food retail outlets.

The elusive nature of food quality is well recognised by consumer groups. Rather than comment on quality directly, consumer groups will address other criteria (often as surrogates of quality) such as safety, price and choice. Where consumer groups would want to go further than retailers is to link choice with knowledge: the informed consumer. As one consumer group official put it: 'we're not there to dictate what consumers should or shouldn't do, should or shouldn't have, but we do believe people should have the information they need to make up their own mind.'

Similarly, within government, the idea of choice is important. Once safety criteria have been satisfied, consumers should have the freedom to make decisions about what they want to buy. For MAFF this involves, firstly, making sure that food purchases 'are not going to make consumers ill or kill them', and secondly, that 'the consumer is not deceived' (interview with official). MAFF is here performing a balancing act between its traditional role of protecting the consumer from dangers with products and a newer role of enabling consumers to make choices. The balance then shifts between government and private sector as to who has responsibility and rights for ensuring food safety. As one senior MAFF official explained:

> We therefore set a legal framework *putting the onus on* the trader, the producer, the distributor, the retailer, the manufacturer, whoever to do certain things so as to make sure that the consumer can make the purchase with confidence that the information given to the consumer at the time of purchase is correct. And that the food is safe. So that's the essential purpose that we're here for.

The shifting balance of rights and responsibilities for food safety means that government is also increasingly engaged in sharing its authority. The implications for the construction of food choice and the relative power of the key actors in that construction are thus undergoing some changes. The case of food irradiation provides an interesting example of the way in which government and supermarkets help to construct consumer food choices. One MAFF official explained the situation as follows:

> We as a department are very, very nervous about legislating to make food irradiation legal. But there were simply no scientific grounds on which you could continue not to permit this process. There were quite a lot of emotional grounds on which not to permit it. But ministers eventually decided that this was no basis on which to restrict consumer choice. The process of irradiation [did not create additional risks], and could arguably be said to have real benefits because all the microbiological risks associated with that food product were eliminated in the process. So the scientific arguments were that this process should be approved immediately. But the fact of the matter is it is not simply being used in this country because the retailers have said that they are not going to handle it. Consumers will take fright you see.

As a result, on this view, choice is diminished. Curiously, in this case the traditional roles of public and private actors have been overturned. The government argues that consumers should have the freedom to consume irradiated food if they wish, while the supermarkets, conscious of their growing responsibilities to protect the consumer interest, are wary of becoming embroiled in potential controversies.

Conclusions

In exploring the regulation of food and consumption rights we have drawn attention to the parts played by key actors: government, retailers and consumer groups. We have shown how their strategies can be interpreted through two competing notions: an individualised *freedom to* consume goods and a collective *freedom from* adverse effects. Over time, as the nature of food supplies and products shift, so the focus of public sector regulation moves, with a relatively greater emphasis on safety and quality issues and less on security of supply. Although these may be competing perspectives, they are not mutually exclusive. Government can have a role in promoting rights to food choices and also to the protection of the consumer. In practice, therefore, as we show in Part III, these perspectives can be found operating alongside one another.

The Conservative governments of the 1980s and 1990s devolved new rights for the private sector to utilise in the form of freedoms to consume. Essentially, the provision of these rights has devolved to the retailers, and only through them to consumers; they particularly take the form of assessments of food quality. Not surprisingly, it is in this area in particular that retailers show themselves to have more sophisticated notions of the consumer than the state. Rights have not resulted from social struggle led by consumer groups but have been bestowed upon consumers through other economic interests. The critical relationship is that between retailers and the state and their ability to regulate food standards and quality. It is therefore to the retailers that we must look to understand the provenance of the new patterns of food rights that have emerged. These rights are sympathetic to their competitive strategies which we outlined in Chapter 3. As we see in summary, in Table 4.2, the evolving changes in government regulatory culture have been matched by a retailer-led individualised consumer culture where the provision of quality choices through competitive retailing has been paramount. The new compatibilities between regulation and consumption have been thus based upon a new partnership of power and responsibility between government agencies and retailers. In this partnership, it has been the retailers who have increasingly formed the means to deliver the rights to consume.

The development of new food rights, however, brings with it new duties (Jones 1994: 16). This in turn means that retailers must be able to exercise considerable influence over those that supply them to ensure the quality of the food that they sell. The ability of the major retailers privately to regulate their supply chains and thus guarantee food standards has been formally recognised with the Food Safety Act (1990). One of the provisions within the Act is that retailers and manufacturers must be able to show that they have exercised due diligence in their activities in case there should be a problem with a product. The Act, however, simply legitimated and reinforced what had become commonplace and made due diligence a legal defence. As a senior figure at one of the major retailers commented, well before the

Table 4.2 Key features of the food consumer as represented by the state and retailers

	State	*Retailer*
Citizenship rights	freedom from	freedom to
Perspectives on choice	collective consumer concerns	individualisation of the consumer
Who represents the consumer?	state	best represented through retailers
Why is the consumer represented?	protection	retailers are harbingers of a) choice b) quality
Food and quality		food commodities become encapsulated in new value and quality criteria
Use regulation to	maintain standards	create competitive space (e.g. innovations in new foods)

Act 'we said to people [i.e. suppliers] ... we expect you to be operating under quality management systems'. By determining quality within the supply chain, the major supermarkets seek to gain competitive advantage.

The much greater diversity of consumption opportunities certainly compared to the late 1940s and 1950s, allied to changing patterns of regulation, means that notions of quality are now increasingly embedded in *where* consumers purchase products. This in turn depends on the ability of retailers and manufacturers to engage in supply chain management, which will vary enormously. There is, therefore, the potential for already significant gradations and hierarchies of food quality between different retail outlets to be further accentuated.

This in turn has significant implications for our understanding of citizenship. Rights, especially socioeconomic rights, of which food is one, are most usually held by all members of a society. When freedom from want was the dominant food right, then it applied to all citizens reasonably equally, although of course some would always have the opportunity to exercise their freedom to purchase whatever foods they wished. In contrast, the freedom to consume, which is becoming increasingly prevalent, is highly differentiated and for some is unlikely to exist. Thus any notion of a collective citizenship right faces challenges from the assertion of individual consumption rights. Ironically, our argument shows how much that is done in the name of the consumer often has a limited consumer input and may indeed benefit consumers differently; but then again, the passive nature of British citizenship is one of its features (see Turner 1990). Moreover, we have also shown that to understand how citizenship rights are formulated and modified through time it is essential, at least in the case of food, to bring together the

state and retailers and link them to regulatory strategies. Given the importance that regulation plays in rights, the changing nature of regulation is
analysed further in Chapter 6. Before that, however, we pursue the theme of
integrating the key actors who construct the food matrix in our analysis of
regulation and retailing. In the following section (Chapters 5–7), we detail
the emergence and operation of this new model of food consumption.

Part II
National strategies

5 Food consumers

The limits of formal and collective representation

Introduction

In the previous chapter, we explored our understanding of citizenship and food rights. Here we seek to progress that analysis by analysing the ways in which consumer interests are organised and represented. In doing so, we go beyond the study of consumer groups to embrace the role of government as well. This is because government has traditionally played an important role as the representative of consumer interests, and, of course, in defining food rights. As we argued in Chapter 2, food policy is increasingly formulated at the European level. This has important implications for consumer groups. Perhaps the most important of these is that, within a European context, food regulation is designed to support the single market and consumer protection is part of ensuring that the market operates smoothly (see also Jukes 1992). Our review of the organisation of consumer groups and the role of government is designed to cast light on the reasons why food consumer groups have long been a marginal presence in British food policy making.

The EU and the consumer

The organisation of the Commission and its ability to be seen to deal more openly with consumer interests is in marked contrast with the way in which British government has traditionally been seen to operate. This takes two forms: first, there is the Commission's organisational structure with DG XXIV's (Consumer Policy and Consumer Health Protection) responsibility for the consumer, and second there are the more general consultative arrangements that exist to gather the views of consumer organisations. It is important to emphasise that the Commission's consumer representation bodies (such as the Consumer Consultative Committee) have caused considerable difficulties for consumer groups in that they traditionally embrace a much wider set of interests, such as trade unions, as part of the broader strategy of ensuring that social interests are represented in Europe. For consumer groups, however, trade unions are too closely associated with producer interests and so may undermine a distinctive consumer message. Similarly, as we shall see, the existence of a DG charged with responsibility for the consumer does not ensure that the consumer gets equal status with producer interests.

Within the Commission, there are three key DGs involved with food and consumer interests, and these have experienced quite different fortunes in recent years. DG III, which has responsibility for Industry and the Single Market, has concentrated its attention on public health (e.g. additives), protection of the consumer interest (e.g. labelling) and control and enforcement provisions. From a DG III perspective, this means that the consumer can legitimately expect from European legislation: (i) that food should be safe and fit for human consumption; (ii) that consumers should be properly informed of what it is that they are buying and what is bought should correspond to the information given; and (iii) there should be a right of redress if either (i) or (ii) are not met. Similar sentiments are to be found in DG XXIV, which, adopting a broader perspective, believes that in a genuinely functioning internal market there is still a need for legislation to protect the consumer; these are ground rules to allow the market to fully function. With specific regard to food, however, a slightly different emphasis begins to emerge. DG XXIV officials believe that when consumers purchase food it should not be unhealthy, it should meet basic standards and should be risk-free. Whereas in DG III, assumptions are made about an informed consumer who should be free to choose, DG XXIV brings to the fore the need for products to be free from risks to the consumer (see Chapter 4 where this distinction is discussed further).

DG III has an Advisory Committee on Foodstuffs, including representatives from five or six interested groups such as retailers, manufacturers and consumers. The committee is consulted on new proposals. Representatives on the committee act in two ways. One of these is as a distributor of information. So, for example, the consumer representative should distribute relevant information and consult with other consumer groups. That consultation will also lead to the second role, which is to act as a representative of consumer views which involves trying to secure broad agreement on issues – a single voice is stronger than many – and thus involves the filtering of consumer group views to DG III.

DG VI, which deals with agriculture, has seen its role weakened in the 1990s. It has lost responsibility for the veterinary services and scientific committees on European food legislation, and thus its role in the implementation of food hygiene legislation, to DG XXIV (Consumer Policy and Consumer Health). This was formerly known by the narrower title of Consumer Policy Service and did not have the status of a DG. This change was a direct result of the BSE crisis and its ramifications for European agriculture and consumers. DG XXIV is therefore amongst the newest of the DGs, and one of the smallest. Before taking on its new competencies it had a staff of 120, but with its new responsibilities it increased its numbers well over threefold to 405. The new rationale for DG XXIV was to try and restore some balance in the marketplace between sellers and consumers, which it was perceived at the political level within the Commission – but not necessarily the member states – had been unfairly tilted towards the former. Not only did DG XXIV's existence depend upon political will, but so too did the reorganisation from which it benefited, which was seen by officials to have come from the very top of the Commission with only minimal involvement from the member states. Thinking within the Commission may have been influenced by the suspended vote of censure which it received from the European Parliament for its handling of the bovine spongiform encephalopathy (BSE) crisis in 1996. Within DG XXIV, the shifting of responsibilities away from DG VI was seen as sending an important signal that agricultural policy makers could no longer be trusted to protect the consumer. There was a strong belief in DG XXIV that DG VI handled BSE in the same way that it handled other farm production issues, and simply did not fully recognise the health implications. As one official put it: 'By removing responsibility from DG VI and placing it with DG XXIV, any thread of doubt about whose interests were being protected was removed'.

With DG XXIV to the fore, in the 1990s a more stringent approach to food hygiene regulation was promoted which attempted to reconstruct consumer trust in European food. Self-regulation was not seen to be a viable option; instead, what was needed was seen to be third party regulation (i.e. by the Commission of national food inspectors), so reassuring consumers. A further aspect of proposed reorganisation will therefore be the creation of a number of posts to strengthen control mechanisms such as more on-the-spot inspections of member states' inspectors, or as an official put it, 'EU control of the controllers', to ensure that member states carry out the necessary number of inspections and adopt correct procedures.

More generally, DG XXIV has two roles: to ensure that consumer views are represented in the EU policy-making process, and to develop consumer representation in member states. For our purposes here, it is the first of these two roles which is the more interesting. What it reveals is the way in which the EU tries to ensure that consumer groups, which are in comparison to industry weakly organised at the European level, do have some opportunity either to directly represent themselves or to have their interests represented by DG XXIV. What is happening is that DG XXIV is acting so as to nurture consumer groups and their expertise, 'getting them involved in the consultation process because it gives them knowledge so they can play a role in public debate'. Until consumer groups are sufficiently proficient at representing the consumer, then DG XXIV must also play a role in acting as the consumer representative in European policy making. That is 'representing the European Consumer interest'. So how might this occur? In the early stages of legislation, there will be considerable negotiation between DGs to arrive, if possible, at an agreed position. Here there is an opportunity for DG XXIV, which looks across other DGs, to make a direct input. When other interested bodies are consulted, then DG XXIV can ensure that other DGs do contact relevant consumer groups and encourage those DGs to do so on a regular basis by establishing appropriate consultative mechanisms. DG XXIV can also act as a post office at which it receives information from other DGs and passes it on to consumer groups, who will send their responses back to DG XXIV to be forwarded to the relevant DG or send it direct to the DG themselves.

Clearly, the role that DG XXIV performs overlaps with the activities of consumer groups. By far the most important consumer group operating at the European level is the *Bureau Européen des union des consommateurs* (European Consumers Organization) (BEUC), which works well with DG XXIV. BEUC is a federation of thirty national consumer bodies drawn from all EU member states and from European Free Trade Area (EFTA) countries and Slovenia. Trying to hold together such a diverse membership is a challenge in itself. The different national consumer bodies have diverse traditions for securing consumer representation, their own specific concerns and varying degrees of authority to speak on behalf of consumers, as some groups will be small while others will employ hundreds of staff. BEUC addresses a range of consumer issues from legal matters, environment, health and food through to safety. Within each of these policy areas, the agenda for BEUC is:

> aimed towards opening the single market to consumers. ... [T]he evolution of policy on the Single Market has focused on the supply side, on the underlying assumption that consumers will benefit. This assumption is ... more or less valid but it is based on a rather limited concept of consumers as passive beneficiaries rather than active participants in the Single Market process. We argue for a more dynamic concept of consumers and for policies which can encourage them to be more active demanders and initiators, actively exercising choice and driving forward the Single Market process as a result.
>
> (BEUC 1996: 5)

BEUC aims to contribute to EU policy in three ways. First, it responds to DG consultation documents, draft directives and so on. As one BEUC lobbyist put it, 'the aim is to work directly with the people who write the proposals'. Second, it participates in the Commission's Consumers Consultative Panel. Third, it lobbies members of the European Parliament. Not surprisingly, BEUC gets the most sympathetic hearing from DG XXIV, but clearly recognises that 'DG XXIV is not as powerful as DG III and DG VI are'. In a significant admission of the weakness of its relationships with DGs, BEUC claims to have 'established particularly good lines of communication with the European Parliament through direct contact with key members and through Parliament Committees'. The influence of the Parliament on proposing and preparing legislation may not be as great as the DGs, but BEUC receives a more sympathetic audience and lower thresholds to be able to contribute to policy developments, so providing it with opportunities that it would not otherwise have to promote consumer concerns.

BEUC's agenda is much wider than food, and so food has to compete for attention with other consumer issues within the organisation. During the 1990s it was possible to discern a steady increase in the prominence accorded to food, which then accelerated rapidly to become the clear priority issue when the potential health implications of BSE became apparent. For example, in the Annual Report for 1992, foodstuffs were a subsection of health, safety and quality issues and covered nine paragraphs, a similar position to that of 1993. In 1994 the report was reorganised and food issues were accorded a separate section and covered thirteen paragraphs; in 1995 this had grown to fifteen paragraphs over five pages (in 1992, they had taken up two pages). By the following year (1996), BSE had become the food priority issue (BEUC 1996).

As a mark of the importance of food to BEUC, food issues are dealt with by a specialist Food Officer. Aside from BSE, BEUC has been fairly consistent in the issues on which it has campaigned. These include the use of food additives, colourings and irradiation of foodstuffs, where it has fought liberalisation on the grounds of protecting consumer safety; labelling, which it wishes to see strengthened to help consumers make more informed choices; and hygiene, where it supports raising standards and the wider application of HACCP. Analysis of Annual Reports and regular submissions to the Presidency makes for something of a sorry story. BEUC positions are almost routinely ignored, or at best only partially met. As an official from DG III commented, the food industry is seen to have much greater resources than consumer groups. Indeed, the food industry has a number of specialist bodies (for example, the *Confédération des industries agro-alimentaires de l'UE* (Confederation of the Food and Drink Industries of the EU) (CIAA), representing food manufacturers, and EuroCommerce, representing retailers) whereas BEUC is a multi-purpose organisation campaigning on a number of fronts. Moreover, as one campaigner argued, a 'high level of expertise is now required to take part in the food regulatory debates' and much of that knowledge is in the hands of producer organisations.

Nevertheless, it would be unwise to simply dismiss BEUC as ineffectual. As an official in DG XXIV argued, during the late 1990s 'the consumer groups' presence exceeds what would be expected by their limited resources. This is in large part because the political pendulum has swung in favour of protecting the consumer and away from protecting the producer'. While an official from DG XXIV may take a more optimistic view of consumer groups than those from other DGs, it does indicate that a shift of influence has recently taken place. Whether this is more than transitory will depend upon the ability of consumer groups to improve their lobbying through greater professionalisation, resourcing and so on, and upon continued political commitment to the consumer.

The role of British consumer groups

Three British consumer groups are members of BEUC. These are the Consumers Association (CA), the National Consumer Council (NCC) and Consumers in Europe Group (CEG). In addition, there are a small number of other consumer groups with an interest in food, including the Food Commission (FC), the National Food Alliance (NFA) and the National Federation of Consumer Groups (NFCG). Below we trace the strategy and tactics of these groups, their links to national government and to one another. The groups themselves adopt different, often complementary strategies. At the peak are the two major consumer organisations, the Consumers' Association (CA) and the National Consumer Council (NCC). Far less well resourced but having a similarly generalist outlook is the Consumers in Europe Group (CEG). Two specialist food groups are the Food Commission (FC) and a coordinating body, the National Food Alliance (NFA). One further coordinating body, but again with a generalist outlook, is the National Federation of Consumer Groups (NFCG). Below are brief descriptions of each of these groups.

Consumers Association

Founded in 1957 to improve the standards of goods and services, the CA trademark is rigorous testing of products and reporting on them in its magazine *Which?* It has a staff of around 500 and is funded largely by individuals subscribing to one or more of its magazines and joining the CA. It has three people working on food and, as elsewhere in the CA, the emphasis is on the scrutiny of products to make sure that they are properly labelled and meet any claims that are made of them, and campaigning for greater information for the informed consumer.

National Consumer Council

The NCC is funded by the Department of Trade and Industry (DTI). It is a campaigning, lobbying and research body. It has three main departments: consumer support, campaigns and policy. It has elucidated seven 'principles' of consumerism by which the effectiveness of supplies of goods or services, public or private, are measured. These are: access, choice, safety, information, equity, redress and representation. The policy department consists of about nine researchers addressing different aspects of consumer policy. Essentially the department's staff spend much of their time producing responses to government consultations.

National Federation of Consumer Groups

This is a small organisation of two and one-half paid staff based in Newcastle.

They exist mostly on two grants, one from the Consumers' Association (CA) and the other from the DTI. The organisation is a federation of sixteen grassroots consumer groups based all over the country (from Aberdeen to Plymouth). It grew out of an initiative by the CA over thirty years ago to deal with local consumer issues. The NFCG essentially has two roles: one is to provide support and training to the local groups, and the other is to represent their opinions nationally to government, retailers and manufacturers. As one of its officers remarked, the NFCG sees itself 'very much as a generalist organisation ... so that food is only one fraction of our total work'. In the early 1990s it set up a food network to provide better information to members and to become more informed about what members thought about food.

Food Commission

Originally known as the London Food Commission, it was created as an initiative of the former Greater London Council (GLC). With the demise of the GLC it dropped 'London' from its title and now relies for its income principally on the sales of its quarterly journal, the *Food Magazine*. It is a small body with about five staff, most of whom are part-time. It has largely kept to its original agenda of exposing the links between food production and consumption. Its activities have shifted somewhat over time, as in its first five years it was mainly concerned with producing factual reports and gaining media coverage for them. Since then, the Food Commission has focused mainly on the media coverage of its campaign work. It is through the media and its journal that the Food Commission tries to influence public opinion.

Consumers in Europe Group

The impetus behind the formation of the CEG in 1978 was Britain's entry into Europe in the 1970s. In particular, the National Consumer Council and Consumers' Association felt the need for a specifically European focused consumer group, a body that could identify the implications of European legislation and policy for British consumers. Its membership is drawn from a broad cross-section of consumer and welfare organisations. Consumers in Europe Group is funded by the DTI through the National Consumer Council, but is an autonomous organisation. Like the National Consumer Council, it covers the range of consumers issues but has for long had a part-time food officer.

National Food Alliance

The NFA was founded in 1985 as an attempt to bring a more co-ordinated approach to food groups and consumer groups interested in food. It is a small group, with a staff of four who survive on 'a wing and a prayer, hand to mouth'. Funding is from a mix of small grants, sales of publications and membership fees. The NFA has up to sixty members drawn from diverse sources, including trade unions, environmental groups and the National Farmers' Union (NFU). Within the NFA the organisation is seen as having two roles: one is to service the membership, and the other to represent the membership at meetings. An important part of its work is information exchange, so all who attend its meetings are given five minutes to talk about their current projects, and 'this is one of the things that people find most valuable'. Resource constraints mean that long-term planning or strategic thinking are not possible. The NFA has a reputation for being one of the more radical food groups (along with the Food Commission). In practice, the picture is more complicated. The group has an 'odd portfolio of projects', some of which are radical while others are straightforward. As its guiding light, Jeanette Longfield, explained, it is an almost inevitable situation 'because we have a broad church of membership, and if we lean too far to the left or right, then we will lose members on the opposite wing. Therefore, we have to strive to preserve this balancing act. I'm not entirely sure it is successful but we have got through so far.' Although the organisation is an alliance, members are not expected to act in a cohesive manner in responding to new or controversial matters. In order to try and ensure that members do stay on board, there can be exhaustive internal consultations on priorities.

The food consumer lobby agenda

Amongst the most notable features of the national British food lobby is its small scale in comparison to food producers, manufacturers and retailers, and its diversity. Given the size and limited resources of most of the consumer groups there is a sense of responsibility that they should cooperate with one another to maximise their potential impact. The extent of, and nature of their links to one another is explored in the following section. Such cooperative working depends upon a number of features, one of which is broad recognition of a common agenda. Here there seem to be two principal forces at work: the desire to create a positive agenda for change, and simply responding to the agenda of others.

Although small, the NFCG is typical of food consumer groups in that it takes a particular interest in food labelling. So, for example with regard to the Bovine Somatotropin (BST) growth hormone, another issue which concerns the National Federation of Consumer Groups, it would wish to see BST-produced milk labelled so that people can decide whether they want to buy it. The basic consumption principle by which it, and virtually all of the other consumer groups, operate is one of informed choice. Expressing sentiments which the other groups would largely share, an official from the National Federation of Consumer Groups said, 'we're not there to dictate what consumers should or shouldn't have, but we do believe that people should have the information they need to make up their own mind.' It, like a number of the other groups, subscribes to the National Consumer Council's 'seven principles of consumerism'.

On the negative side, in determining their agendas the small generalist groups face a common problem. As one put it: 'To a large extent, I think it [our agenda] is reactive.' The key constraint is that 'we don't have the resources to do very much active in the ... [food] fields'. Thus groups can seek common cause in coming together in reaction to the proposals of government or the food industry. There are, though, significant exceptions. For example, the National Consumer Council is sufficiently well-resourced that, although its agenda is partly set by the need to respond to government ideas, the researchers are sufficiently immersed in their specialist areas that they can identify up and coming issues which they think the National Consumer Council should address. Interestingly, those operating within the food lobby did not see that the National Consumer Council's ability to create a positive agenda for itself would lead to a situation in which it could unite other groups behind it. This is revealing of the nuances of the different groups and the need to retain their individuality to survive.

One aspect where it is possible to distinguish most clearly the positions of different food groups is in relation to social issues. All the groups recognise that there is a welfare dimension to food policy but the National Food Alliance, the National Federation of Consumer Groups and the Food Commission are most sympathetic to the plight of the disadvantaged consumer, followed by the Consumers in Europe Group and to a lesser extent by the National Consumer Council.

Links between groups

One byproduct of the small scale of the specialist food groups and the small number of staff dealing with food in generalist consumer groups is that they enjoy close links with one another at both a personal and organisational level. Staff are members of different groups; for example the National Federation of Consumer Groups is a member of the National Food Alliance and receives funding from the Consumers' Association, and staff regularly meet either in those groups or on government committees. Here there exists a community of professionals who are highly committed and in some cases very highly respected. In a small network, reputation matters a great deal. Relationships of trust between the groups are for the most part very high. This has an important consequence in that groups will, at least informally, divide their lobbying and policy work up amongst themselves. As one group leader put it:

> Generally what tends to happen is we will meet with the Consumers' Association, the Consumers in Europe Group, maybe the National Food Alliance, and talk about all the work programmes and what people are

doing. I'm particularly careful not to be doing the same work as some-body else. There's just absolutely no point.

Here group leaders may be prepared to subsume the interests of their organ-isation for the benefits of cooperative working for the benefits of the food consumer lobby as a whole. Perhaps the one area where one domestic group is seen to be in the lead is the Consumers in Europe Group in relation to its European work.

Amongst the consumer groups there is, however, perceived to be a tension between the Consumers' Association and the National Consumer Council. As a member of staff in one of the groups put it: 'there is some sort of competition perhaps for headlines. Not really at the level of working staff. I think that staff work well together.' Part of the problem is that it is in the nature of the National Consumer Council's approach to sometimes gain wide media coverage, because it only does policy work and its responses to government can attract headlines. The Consumers' Association, on the other hand, is more market-led and is more in need of the headlines to maintain its profile and ultimately its membership.

Links to government

The groups themselves have quite different strategies towards engaging with government. On the one side is the National Consumer Council, which has well-established links which it wishes to develop further, and on the other is the Food Commission, which is wary of becoming too closely involved with government. As we shall see, the range of strategies appears to be partly a tactic adopted by the groups themselves to try and maximise their opportu-nities for influencing government.

One of the most important avenues for groups to obtain access to govern-ment is through MAFF's consumer panel. As one group leader commented, 'that's a very good channel'. The formation of a consumer panel was some-thing of a novelty and treated with some suspicion to begin with:

> Well, in common with the other consumer organisations, we were extremely sceptical when it was set up. It was set up in the light of what Edwina Currie said about [salmonella in] eggs. They'd [government] got to be seen to be doing something. I think they've largely won us over, in that it isn't a hollow PR exercise, and they have moved quite considerably on a lot of things that the consumer panel had been pressing for.

There is a clear sense among consumer group leaders that their influence within government has grown since the late 1980s, 'partly through the setting up of the consumer panel, partly because we have quite a lot of people now on MAFF committees. So we feel we're representing the consumer in those places.' It is important to remember, however, that access does not mean influence and that consumer groups have started from a very low base of access to government. It is therefore important not to overemphasise what has happened. (A good review of the influence of consumer groups on MAFF is in Consumer Congress and National Consumer Council (1994).)

In any case, the groups themselves have realised that there are considerable benefits to government and dangers to their own positions as a result of the consumer panel and groups such as the DoH Food Authenticity Working Group. The concerns of the consumer groups are twofold. One of these is that government has appointed non-consumer group experts to its consumer committees, so diffusing a specialist consumer input. Such people are appointed as individuals, do not report to any organisation and cannot call upon the resources of an organisation to support their work. It is not necessarily that such people are more compliant or less informed, but they are removed from the food consumer lobby. One MAFF official adopted a much more jaundiced view:

> But how representative are consumer organisations? I have to say I am highly sceptical of them. They are mostly staffed by middle class do-gooders who feel that they are representative of the nation as a whole. Quite frankly they are not. And to what extent they actually canvass consumer opinion before they express a view – well, I doubt very much whether there is any process of consultation.

The other consumer group complaint is that whilst there are a number of committees now in operation which do have a consumer input they operate in a somewhat disparate manner. There is no coherent consumer agenda from government, and it is difficult for groups to put one forward when their knowledge and expertise is dispersed.

One group that has sought to keep its distance from government is the Food Commission. It tries to adopt the classic outsider group position, using its high profile to make good use of the media and targeting wider public opinion rather than simply policy makers. As one of its leaders commented:

> We're being increasingly consulted even though our resources are becoming ever more fragile ... I would say that we are always aware of the potential for being incorporated and although we have been incorporated increasingly, the danger is that our input will be influenced that way, and we have to keep reminding ourselves of where we stand, and not to be incorporated and hence subsumed and lost.

As part of the ongoing process of incorporation, staff at the Food Commission have been invited to attend discussion at MAFF and the DoH, but are more usually asked in their individual capacity rather than as staff members of the Food Commission to serve on consultative groups.

Other groups have been less concerned about incorporation than the Food Commission. For example, the Consumers in Europe Group have reasonable informal contacts with some civil servants and ministers. The Consumers in Europe Group have a good reputation. They have a member on the MAFF consumer panel and another on the Food Advisory Committee. Meanwhile, the National Food Alliance sees one of its tasks as being to act as a lobbying organisation. According to its leadership, its strength as a lobbyist is its diverse membership making it less easy to dismiss, and providing the opportunity for additional credibility. As it was explained: 'if, for example, the NFU [a member body] signs up to a proposal then people respond with "oh well, there must be something in it".' Although the National Food Alliance receives material from MAFF, it rarely has the resources to respond. It therefore tends to act as a post office and pass material on to relevant members:

> Usually the only time that the Alliance responds to a consultation document as the National Food Alliance is if one or more of the members says that there is something screamingly important. They say that they do not want to put one [a response] in on their own but please can there be a National Food Alliance response.

Although the National Food Alliance may try to portray itself as a broad coalition others within the food lobby see it and its role somewhat differently. As an official within the National Consumer Council explained:

> I think the National Consumer Council is seen to be the mediator or the sensible voice. And then on the one extreme the National Food Alliance is seen to be the extreme end of the food lobby, and perhaps the Food Commission are nearer to them. There seems to be a perception by government that these food groups are a little bit extreme. But of course, I think they all have a role in the framework. If they weren't there to be extreme, then when we took the middle position, which might be quite a radical one anyway, if there wasn't someone more radical than us, then we wouldn't be accepted to say what we say.

Here, then is a sense that the lobby needs a spread of groups to actually make itself heard and that the groups have an interest in maintaining that diversity to best represent the consumer to government. It is partly in their agendas that the groups will differ but, perhaps more importantly, they will also differ in their tactics. Again, from the moderate position of the National Consumer Council, the National Food Alliance and the Food Commission were seen to:

> have different tools than we do. They use the media, they use the shock-horror approach. That's the only tools that they have. Because we have built up the relationship with government departments, we have built

up the trust of government in that all our work is usually based around sound research – whether its academic research, whether its in-house research, and our arguments are therefore more based on something we can prove. It's much more acceptable and we just have this position within the policy framework.

Once again, however, it would be easy to overestimate the influence of consumer groups within government. As an official within a major consumer group commented:

I do feel that retailers and manufacturers do have a lot of weight ... the thing to remember is that a consumer organisation has one person working in the whole area of food, the whole area of utilities, the whole area of international trade – that is a minimal resource compared to an organisation like [a major retailer].

The official continued:

You know, the industry lobby is just so, so much greater. The only weapon that a consumer movement or organisation has is that of opening up processes. And demanding to take part at the level of decision making, because we don't have the same resources. We don't have the same resources to put into it. So we can only demand a sort of opening up of the process as we go along and to be present at the table.

Such ambitions are modest, and if the experience of the consumer panel is replicated elsewhere it may be a double edged sword: offering access but diffusing influence.

Conclusions

At the European level, quite clearly the ability of DG XXIV to involve itself in contributing a consumer viewpoint in EU policy making is a reflection of the weakness of consumer groups. A danger of trying to nurture consumer pressure groups is that they fall into a dependent relationship with DG XXIV, on which they rely for access to other DGs (thus stunting the development of their own lobbying skills) and for information on EU policy making. The end result is likely to be the development of a clientelistic relationship between DG XXIV and consumer groups. At the same time, DG XXIV runs the risk of simply being seen as an extension of consumer lobby groups by other DGs and undermining its own credibility.

Nevertheless, there are signs that the consumer is an important political concern in Brussels, and the rising power of DG XXIV is but one indication. Another is that organisational reform at the European level to try and restore consumer confidence in food products has been quicker and more radical than that undertaken in Britain. The former Conservative government tried to minimise reforms to MAFF. The Labour government's proposed Food Standards Agency may help to restore consumer confidence and to offer a greater consumer input into food regulation and policy making.

Perhaps the best indication of the marginalisation of consumer groups in British policy making is that Europe has offered new opportunities for them to get their voice heard. In the machinations of European policy making, a classic strategy for pressure groups is to convince national government that they should promote their case in Brussels. There is no indication whatsoever that consumer groups have been able to convince MAFF of the validity of their views and to advance them at a European level. For most food consumer groups, intensive lobbying in Brussels is not feasible because of the cost. Instead, they find it much more efficient to cooperate with Europe-wide bodies such as BEUC and 'to try and influence staff members in BEUC to take up an issue and follow it through themselves'. Indeed, there is something of a conundrum here: food safety issues have reached new heights of public concern, but this has not led to a corresponding growth in the size or influence of food pressure groups. Given the widespread misgivings about the former Conservative government's ability to ensure that food was safe, such a situation is even more perplexing. Part of the reason why public concern has not fed through into food pressure group membership (for example, in comparison with broader environmental groups) is that retailers have played a key role in assuaging the public concern. As we have argued, corporate retailers are increasingly the food safety guardians of much of the food that we buy. Given the hierarchy of retailing that we outlined in Chapter 3 and its implications for projecting hierarchies of food quality (developed in Chapter 6), then precisely those people who are likely to be most concerned about the quality of their food have the freedom to purchase it from the retail outlet of their choice. Thus food quality, rather than becoming an issue around which the consumer collectively organises, is diffused amongst the aisles of the major retailers. How retailers achieve this is examined next.

6   The retailers

The emergence of retailer-led food governance

Introduction: re-regulating the British food sector

The growth in the role of corporate retailers inside as well as outside government has been one expression of the developing private-interest model of regulation, whereby significant former government functions are bestowed, for what are seen as more effective management, to private sector interests. New forms of private-interest regulation have been developing in the British case both as a cause and a consequence of the growth and maintenance of retail power. It is important to realise that such a model does not deny the continuing role of government in food regulation; rather, it changes its shape. From the tried and tested macro- and meso-corporatist forms of regulation, which have been so much a part of the postwar agricultural and food 'compromise' in Britain, the past decade has witnessed a significantly different policy regime. While the traditional corporatist arrangements have tended to wither rather than be eradicated completely, we have seen the progressive development of new micro-corporatist relationships developing despite, rather than because of, the existence of established structures of mediation. This has meant that the nature of food regulation in Britain has become increasingly dense, just at a time when the rhetoric of deregulation was at its peak.

Regulatory policies and the structures in which private interests are empowered involve making retailer interests legally co-responsible, or sometimes even completely responsible, for the implementation of activities that would otherwise have been a state responsibility. This happened most notably with the Food Safety Act 1990, and more recently with the passage of the Food Hygiene Directive 1995. The broader shifts in the political economy of the nation state, and the increasing 'consumer-led' transfer of power down the food chain to retailers (see also Doel 1996; Marsden and Wrigley 1996) have combined to empower retailers as harbingers of public policy through private means. Such shifts in responsibility are based on new and interesting combinations of economic and political power under conditions where consumption and food quality issues, and particularly their articulation and representation, are central for the maintenance of both. Whereas the dominance of a state-centred model of regulation is often seen as necessary because of the lack of public confidence potentially generated by 'the bazaar of the market' and the 'tragedy of the commons' scenarios that self-regulation can often suggest, retailers and the state have evolved working relationships which maintain public legitimacy and market power through a coming together of their interest in privately and publicly needing to demonstrate their mutual role in serving the 'consumer interest'.

Indeed, the newer phase of flexible micro-corporatism is increasingly concerned with dealing with the legitimation problems of consumers rather than producers. As a result, issues of 'quality' rather than quantitative supply become both more immediately significant and pressing for the maintenance of retailers' competitive space. Moreover, as we will attempt to show, one basis of this changing context concerns the construction of authority and legitimacy of food quality and 'the consumer interest'. These become the substance of micro-corporatist relations, and are increasingly less hard and fast concepts of regulatory activity. Rather, they are continually contested and represented in the range of different regulatory domains which cut across the food policy spectrum.

The making and remaking of competitive space

In the second part of the chapter we examine some of our evidence of the content of these private interest relations. We wish to explain and explore more fully how the contested regulatory domain of food quality is played out with reference to empirical evidence collected from the variety of key agencies involved, particularly retailers, state officials and representative organisations. This evidence demonstrates the highly dynamic and fluid nature of the regulatory domain in which retailers operate. We do so by exploring the contingent and contested nature of regulation, and how retailers' and state interests shape that regulation. In addition, we demonstrate how the shaping of policy by these interests influences the maintenance – at least in the medium term – of retailers' competitive space, both in a sectoral and spatial sense. As we outlined in Chapter 3, by sectoral here we mean the intra-competition within groups of retailers. In the UK this involves in general terms three tiers; the 'Big 5' (Tesco, Sainsbury's, Safeway/Argyll, ASDA, Gateway/Somerfield), the discounter sector and the smaller and more numerous independent sector. This intra-sectoral competition is also played out spatially in the siting and local market management of stores. This is not just an issue of competitive siting, that is, the popularised 'store wars'; it also concerns the social and economic maintenance of local customer loyalties, which in turn are based on dynamic and spatially variable retail strategies concerning food choices and quality (see Chapter 8 on local retail spaces and hierarchies).

By the mid-1990s, the regulatory domains in which the large retailers had played such a part for at least the previous decade became simultaneously more unstable for a variety of market-based and consumer-based reasons. In the regulatory domains (outlined in Chapter 3), public pressures to curb retail prices and levels of concentration have been compounded by the disproportionate competitive effects of the arrival of the new discounters, creating the need for the big retailers to reconsider their 'value lines' and their locational strategies. Public concern over the quality and the origins of foods has also continued. This has meant a heightened degree of activity for the retailers in the regulatory domains, not least in the areas of food quality regulation. As Wrigley (1995: 23) has argued:

> the problem facing the major UK retailers is how to sustain opportunities for capital investment during a period (the later 1990s) in which the avenues utilised during the 'golden age' of the 1980s/early 1990s are likely to be far less rewarding ... Such opportunities are essential to firms which, despite intensifying levels of competition in the UK, remain immensely profitable and have huge cash flows from multi-billion turnovers. Given that the essence of the accumulation process in retailing is the constant need to 'ground' retail capital (indeed, the associated vulnerabilities which flow from that 'fixing in place', and the tensions and contradictions which are played out in the process, form a central component of the corporate strategies of major retailers), strategic diversification to reduce dependence on capital investment solely within UK food retailing has become paramount to the major retailers.

Our research evidence suggests that this 'grounding' also has to take place socially and politically in the form of active participation and contestation in the key regulatory domains outlined. The following discussion details this by examining some of our qualitative evidence concerning retailer strategies and food quality concerns. Within this domain, by the mid-1990s, we can see strong attempts by the retailers to influence and react to a variety of regulatory issues dealing with both the formation and implementation of food quality policy. Some of these deal with statutory and EU legislation (for example, the Food Safety Act 1990 provisions, or the more recent EU Directive on the hygiene of foodstuffs (Council Directive 93/43/EEC)), while others are more generally dealing with the creation and coordination of policy, associated, for instance, with health and nutrition, the government's deregulation initiative or the manipulation of foods (for example, irradiation, temperature controls and genetic modification). Throughout we emphasise the variability between the different tiers of retailing. This variability is both a cause and consequence of the 'between firm' (intra-sectoral) intense competitive relations which are exhibited in the retail sector. These competitive spaces are built and maintained by regular participation, on the part of retailers with government.

The drift of retailers' participation with government occurs both at an individual firm level, that is, between specific retailers and state agencies, and in a flexible and collective way with groups of retailers and government. Sometimes, the British Retail Consortium (BRC, the retailers' trade organisation) becomes the mediator, while on other occasions, and sometimes simultaneously, retailers operate on their own, attempting to create their own competitive space from inside the regulatory domains which affect them. The corporate retailers have to keep one eye on their competitors and one on the government, and at the same time they have to project what seem to be valid conceptions of the 'consumer interest' to government in order for their voices to be heard and legitimised. As a Director of one of the 'big 5' argued:

> We have said to government, for heaven's sake, is the benefit to the consumer proportional to the burden on industry? Ask yourself that question when you are about to regulate. Talk to the people in the industry about what the burden will be before you regulate. Talk to the consumers about what the benefit is actually going to be before you regulate. Don't just throw a Directive into a dark room full of government lawyers, and then put up regulations willy-nilly, which is how it has been done ... Now we are very much at the heart of the deregulation initiative. I actually had people sitting on task forces. The food and drink task force actually addressed the process of food regulation, as well as saying, 'what can we do or undo?' I insisted that they actually addressed the process of future regulation, so that in ten years time, we don't have to sit down again and address another deregulation initiative. We have got this far, we have drawn a line in the sand, and said that anything we do from now on should be done in a consultative way when the burden/benefit ratio is assessed. There is now a total cost–compliance assessment, the need is assessed. There is full consultation, and regulation is introduced only as an absolute necessity. Now, this is the agenda that the government has agreed to work to in the future.

Such powerful negotiating activity within government is not, however, just associated with the literal forms of deregulation. Indeed, it is much more positive in shaping new forms of re-regulation. The job is not to sweep away everything; rather, it is to recreate policy around different power interests and different knowledges and representations of the consumer interest. For instance, as another executive from one of the 'big' retailers argued:

> Lobbying has become much more sophisticated. I mean the big impetus from us was undoubtedly the fact that our chairman was involved directly with the de-reg initiative, that that created a focus in the business. I think the Hygiene Directive is an example of deregulation. We have moved things back to having responsibility within factories for staffing levels, so that you haven't got a tier of people doing the job.

You haven't got inspectors. So I think there are good examples around where we have deregulated. For instance, we were particularly interested in the Meat Hygiene Directive; as a company we did lobby these separately to other retailers. In fact we believe we had a great influence there, and did a good service to the industry.

Similarly, the EU Directive on the hygiene of foodstuffs (Council Directive 93/43/EEC), which came into force in the UK in September 1995, has embodied many of the self-regulatory principles advocated by these 'big' retailers. The retailers, having first developed regulatory systems of supply chain management, have then been promoting them inside the state. In this sense, while the British government has advocated that regulation in the food sector should take place only as a last resort (that is, the 'regulating last' principle), for the retailers, regulation is seen to be more effective when it is constructed 'later'; that is, after it has been developed and applied in the private sector. This 'regulating later' principle is also promoted by the British Retail Consortium.

Another senior retail figure from a somewhat different perspective argued:

> Deregulation is difficult. I prefer to work with legislation actually. Because you've got the guidelines well developed and you know exactly where you are. Deregulation becomes more difficult in the long-term sense. It means that it passes the control very much more back to us, to implement and make sure that it's there. We have our own standards – regardless of government legislation – we've got our own standards. And some of ours are very much more strict than the government's ... The EHOs are looking at a minimum standard. I would say that our standard is significantly higher. There needs to be a constant evolution and improvement. We would look at it that we're never satisfied with what we have got now.

Such sentiments were well understood by senior civil servants, as one from the DTI explained:

> It's not a simple world. It's not industry telling us to sweep away regulation and we do it. One of the things that we think is of great value with the task forces is to take the voice of business into the decision-making process. Not afterwards when the superb civil servant drafted consultation document is issued; with everyone given four weeks to comment and then essentially through the trade association, but rather, to bring business into the discussion at the point at which policy is being formulated. Again, it is not so much the removal of regulation, it's not so much the prevention of regulation, it's getting the regulation

right ... let's ask people about the assumptions that we are making about the impact on industry.

Not all of the corporate retailers are as proactive, as the evidence thus far suggests. For instance, Gateway and ASDA put more reliance upon collective forms of engagement (such as through the British Retail Consortium (BRC) and the Institute of Grocery Distribution). The discounters seem to have little contact with the ministries. In this sense, the competitive spaces which exist between the three tiers of British retailing – the Big Five, the discounters and the independents – are actively reinforced in their differentiated relationships with state agencies and their respective regulatory domains. In turn, re-regulation in the food quality domain at least gives advantages to the Big Five, suggesting that policy formulation and implementation 'trickles down' from them. The retailers need forms of regulation which reflect and sustain them both as a group and in their individual competitive positions *vis-à-vis* each other. As one executive argued:

> We respond to the multiples. We characterise them like this. The multiples are Sainsbury's, Safeway, Tesco, Asda and ourselves. Then you have the discounters – KwikSave, Aldi, Netto, Costco, etc, it isn't quite as simple as that actually, as there are regional discount operators as well. All the multiples are saying to themselves, and to one another, 'we can't afford a gap to exist between the discounters and the multiples'. As a result, it is the multiples, as a body who are bringing down their prices rather than any one supermarket acting in isolation. The supermarket chains do not operate in isolation. It's a very close market in the sense that if we bring down the price of sugar, then Sainsbury's and Tesco will bring down the price of sugar. That's the way the industry works, it is not because there is secret whispering going on. We just watch each other's prices ... Our benchmark is saying that, on our own label, we have to have a product that is among the top three in its field. In terms of product development, we tend to be a follower rather than a leader, given the nature of our business.

We can see from the above discussion that the regulatory culture that typifies state–retailer relations in Britain from the 1980s onwards seriously questions our traditional and dichotomous notions of state or private-sector based models. The time-specific webs of connection on a whole range of food quality policy issues develop new hybrid forms of regulatory practice and formation. These are situated, and derive considerable potency from, a regulatory culture which assumes that the public interest is best served through private sector forms of regulation and government practices, and which depend upon private interests to help to originate the very content of new legislation. This regulatory culture can also seemingly, at least in some cases, absorb policy directives from Brussels and shape them in forms which appease private interests. Indeed, it may be that the retailers and government can increasingly see mutual benefits in lobbying Brussels at the policy formation stage, given the success of the development and implementation of the Food Hygiene Directive.

The changing nature of the regulatory culture described here also has to be linked to the changing significance of consumer culture. Our evidence begins to suggest that the regulatory systems developing are not disassociated from the growth during the past decade of an individualised consumer culture, which has been partly shaped by the corporate retailers themselves (see Pred 1996; Glennie and Thrift 1996). The 'reflexive turn' by consumers, and the individuation and identity politics involved in the consumption process (see Miller 1995), tend to reinforce the political power of corporate retailers in their ability to 'represent the consumer'. As we argued in Chapter 3, in our interviews with government officials it was clear that the material fact that millions of people pass through the main retailers every day gave the corporate retailers an authority in Whitehall which far outweighed the representations of consumer organisations or public bodies. Thus we have to recognise that the marginalisation of representative consumer interests outlined in Chapter 5, is not simply of their own making. The retailers' pivotal position as consumer gateways and social barometers, and particularly their considerable intelligence-gathering activities about consumers, become a powerful representational tool in their dealings with the state.

There are, then, some important interactions between the evolving regulatory culture and consumer culture. The retailers are committed, for their own survival, to promote the constant and dynamic individuation of 'the consumer' through innovating and providing new 'quality' choices. This is by no means inevitable, and has to be constantly constructed not least around increasingly subtle modifications of quality definitions (for example, 'novel' and functional food developments). Consumers have to be encouraged to be progressive in their buying patterns, to try something new and to assess relative forms of food quality. In addition, they have to be captured and persuaded to become loyal. These processes of constantly feeding particular types of individualised consumer culture not only stand to empower the corporate retailer in the local marketplace, they also bestow a growing socially and politically embedded custodial role on retailers which can then be used in policy-making circles. This increases the significance of consumption in policy making and the role of retailers in representing – in increasingly sophisticated ways – consumers and the consumer interest as 'the public interest'. For example, the reluctance of the main retailers to introduce irradiated food products and genetically modified food materials, despite MAFF scientists' approval, demonstrates the ways in which retailers are sensitised to consumer confidence and actions and the need to be seen as responsible custodians of individualised consumer culture. In order for these particular facets of consumer culture to be maintained, government and the retailers have to project themselves as socially progressive. The growing provision of food choice and the elaboration of different degrees and definitions of quality feed this progressive ideology both inside and outside government circles. In addition, such notions of quality hold the secondary effect of creating a widening gap between old style baseline food quality standards (still enforced where necessary by MAFF and the Department of Health), and newer privately regulated quality criteria which lie above absolute quality standards, supplying more choice and variety for 'the consumer'. Retailers thus come to be seen as the harbingers of choice and quality. Nevertheless, these definitions need to be constantly replenished in order for the progressive element to be maintained.

Our interviews reveal how retailers differentially construct quality definitions and project these as legitimate and powerful knowledges in the marketplace. Government officials, while they may rely upon the scientific evidence and adjudications on their committees (i.e. the Food Advisory Committee and the Advisory Committee on Novel Foods and Processes), realise that it is up to the retailers to decide what to sell and how to sell it. One government official explained that the regulation of food safety and quality are seen as broad parameters which the corporate sector can reshape:

> I mean my own view is that it is the retailers who have enormous investment in quality, enormous investment in seeing that their goods are extremely good value for money, extremely good for you, you know all extremely clean and all that. So it is consistent with the government's philosophy, that apart from the food safety regulations – apart from that – we are very happy to see the retailers as the creators of standards in food distribution.

The 'Big Three' retailers are seen both by government officials and by lower-tier retailers as setting the pace on food quality and delivery. As one of them commented:

> Well our profile, I suppose, would be A, B, C and C2s, we have to show, I suppose towards the higher end of the market. You can't be a mass market retailer without being you know, having a very wide spread of consumers. But you do try I think within our business we would say that we are more biased towards the upper end of the demographic scale than anywhere else.

He continued:

> We are continually bench-marking ourselves against the best of the market – and we use the word quality-value. If you take an area that I am familiar with – mangoes for instance – we would import mangoes five years ago and they may not have tasted very good. Well the objective is now that we want very good flavoured mangoes, which are already

ripened and ready for the customer to eat on the shelf. And that would be a measured quality parameter. And there would be a specification for every product. Then we would have had a quality audit, whereby we would say 'is this still the best?'

Through the innovation of over a thousand new products each year and in conducting over 2,500 panels with consumers, the construction of quality is a major aspect of the dynamism of the corporate retail firm. As he continues:

We debate it at length. Then we ask customers to define what they think is quality, they find it very difficult. When we talk about quality of a product we normally talk about its eating qualities. When we ask customers what they think about quality and they spend two hours – and quality to them means the atmosphere of the store, the quality of the trolleys, the quality of their experience when they go in; as well as the quality at the end of the day of their products, and how they eat when they take them home. And all that is wrapped up in what they believe to be quality of their product. So it's impossible to identify one aspect of quality, or to say quality is only when you put it in the oven, and take it out and eat it. Because, before you've got to that stage, all the other things have impacted on it. It does make a big difference if they have had a particularly pleasurable experience when they bought the product, then the quality of that product may not suffer as a result. Not because of the actual quality of the ingredients or how it performs.

Such composite definitions of quality relations between retailers and customers are seen to be different among the lower tiers of retailing. The discounters are seen as offering a different quality–price relationship to their customers:

They offer something different. They offer competitive prices on certain lines that people may be prepared to sacrifice quality to get price for ... So if you're buying something like sugar, or skimmed milk powder, or tin foil you may not expect or may not want quality of that product to perform to a certain level. So you may be willing to sacrifice quality for price. And if they can buy something which they believe gives them acceptable quality, and a very competitive price then they may be prepared to go to Aldi, Netto or Kwik Save to purchase it.

For the market leaders, however, it is a rather different situation, as a senior figure in one leading retailer argued:

The two principal areas for us are product development and own label. These are the two main drivers ... It's about confidence that we have got control of in the products we sell. Our own label products that we have put a lot of work into, protecting the consumer and food safety. And a lot of research has been done in terms of the way people perceive messages. There's a fairly high level of confidence. That if we say something about our products, then obviously we really do believe in them.

But if you take examples like pesticides: we've done a communication there as a quiet revolution, about pesticide reduction. That is a communication direct to our customers, in easy to read format, saying that there are other ways of controlling pests and diseases in crops, instead of putting chemicals on them. When we do put chemicals on them, then obviously we follow the legislation. We test to make sure they are at the right levels.

And it's got to be said, I think, as a company we are much more I mean we are ahead, in terms of being proactive, on these issues. Like animal welfare – a topical issue at the moment – where we have been working for many years ahead of what's happening now with our partnership in livestock scheme ... With our farmers, developing a positive approach to better animal welfare. But a lot of these things take a lot of time to actually implement. So it's a slow process of encouraging people and getting them to go in the right direction.

For the second and third tier of retailers a less composite and proactive set of issues come to dominate. This is illustrated by the following quotation:

The critical issue is what sells. I mean we are in a repeat market. Our customers visit us week by week. Our benchmark of quality is the grounds that we have built up over the years. We will technically specify individual standards for particular products, but that is not against some sort of abstract quality definition, I mean the thing that we try and do is to express our aims in Daily Mirror language because that is what our customers like. We aim to keep the business as simple as possible. What we are in the business of doing is providing a mix of food prices which is what our customers want. We land up with very competitive priced wholesome food, so normally, the best way of working out whether the quality is right is putting it out on the shelves, and see what happens. I mean this is what 'Marks and Spencer' do, they are far and away the best people in the market in terms of new food products. They put them straight on the shelves. They are very intolerant of the things that aren't selling.

In the first place, the own brand came in when the business was trying to become a pale imitation of Sainsbury's, so you have got from a standing start a huge amount of products coming in at a rate of knots. A lot of products went out of variable quality. What we have done in the last two years is to re-examine every single product. We have put them through a detailed analysis and consumer testing, and when we were not happy with the product, we cleared them out. We got to a smaller hard core of product that we were comfortable with where people were very happy with the quality of our product. We were then able to put discount on our flagship brands ... The objective therefore is acceptable quality at outstanding price. This is obviously helping us to protect our underbelly against the discounters. So there is no advantage to go to the discounters. You look at our price – our own pricing – our own label, and there is no reason to go to a discounter unless there is one just around the corner.

As another second tier retailer attempting to claim the high street territory claimed:

Your grandmother is our customer ... They come in because they want a chat. They come in the same time each day, because they know their friends will be there. We see ourselves as creating and strengthening that niche which says 'the High Street frequent shop', and this is the place to come for that. So there is a strong fresh points participation because fresh foods are the sort of thing that you want to buy more regularly. We are catering for a greater sense of impulse buying, because you have people coming in probably four or five times a week.

For the discounters the challenge is to make significant inroads into the richer consumer markets:

I think it has changed just in the four years that we have been operating in this country. We started with the poorest part of the population and now we're moving up and a so-called better class of customers will now shop here. All the discounters will now because people are finding out it's the same product at the end of the day, so why pay more? And the difference that they save might as well go and spend in Sainsbury's on some more wine or whatever.

When we opened on the first day, the customers would take whatever down from the shelves and give it a try ... I think that price is the main weapon in marketing terms so if it is cheap enough people will try it. Because for 7p for a tin of baked beans they're going to give it a try aren't they!

I think, in general, our pricing philosophy is simply to be sold well priced, low priced. So if people have an interest in saving money they will find products that they don't like as much as others and they will say 'these corn flakes are not as good as Kellogg's'; or 'I don't like them for some reason'. Because we have letters in when we change products – and people write in and say 'oh the corn flakes must have changed because they are not as good as they were before'! Any new line has to go through a tasting panel, of which I'm one of the members. We taste everything – against the brands, against the greater multiples' own labels – and so, if it's all gone through a tasting panel, we reject a lot during tasting sessions.

Conclusions

The latter part of this chapter demonstrates how contested and differentiated notions, concepts and regulations surrounding food quality become a major organising concept around which the social, political and economic activities of retailers seek to influence food supply and consumption. As some of the social economists have explored more generally (see Granovetter 1992; Allaire and Boyer 1995), the constructions and representations of quality become a key axis for structuring competition and locating firms at particular sites in these competitive terrains. The competitive spaces that British corporate retailers occupy, as we outlined in the first part of the chapter, are sustained by constructing and articulating conceptions of the 'consumer' and food quality towards both the public on the one hand, and government officials and agencies on the other. In Britain, at least, the uneven development of corporate retailing is assisted by both socially constructed but differentiated notions of food quality, and their use and articulation in systems of micro-corporatist policy making.

In terms of policy, therefore, the conflation of regulatory culture with a particularly dominant, individualised and differentiated form of consumer culture becomes an important basis for maintaining and developing the micro-corporatist systems of policy making in the British food sector. It is on the basis of these that knowledge, authority and power in the retail sector are maintained and developed, and more specifically, the relative competitive space of retailers is structured. For social science scholars, an exploration of these issues starts to demonstrate how aspects of culture and polity link together to provide the ground rules for capital accumulation and structuring the 'situations of exchange' (Appadurai 1986) in the retail sector.

7 Evolving models of food regulation

Introduction

Despite the emergent power of corporate retailers in fuelling authoritative hierarchical and individualised notions of consumption culture, food policy and regulation, the 1980s and 1990s have been dogged by a series of controversies. For the most part these have concerned issues of safety (e.g. salmonella, *Escherichia coli*, BSE) and have raised searching questions about the way in which food is regulated and who undertakes that regulation. One of the many high points of political and public concern occurred in the late 1990s, when within the same month (April 1997) eminent commentators were reporting for the government on *E. coli* (The Pennington Group 1997) and for the then Labour opposition on a Food Standards Agency (James 1997). Both reports exposed the changing nature of food regulation processes, highlighted weaknesses in current implementation practices and pointed to the confusing range of responsibilities and professionals involved in food regulation. Whereas both reports concerned themselves with a range of food regulators and regulatory activities, both pinpointed the key role of environmental health and trading standards departments in ensuring food standards and safety (James 1997: 11).

The deepening crisis in food demonstrated that government and public-interest regulation was still needed, despite the significance of private interests. We too have concentrated our attention on these professionals (see Part III), but have an additional reason for doing so. As our earlier chapters have argued, our interest is largely in the activities of the corporate retailers and how they relate to government regulation and the consumer. The nexus of that relationship is in the store or outlet, and here EHOs and TSOs, together with retailers and consumers, are to the fore. It is here in the micro-situation where they are directly related.

The work of EHOs and TSOs is analysed in Part III, which explores in detail the regulatory practices of local officials. In practice, the boundaries between the work of EHOs and TSOs in relation to food regulation can be fairly fluid. In theory, the distinction between the two professional groups is clear: EHOs are concerned with the protection of the consumer from foods that are hazardous to health through the enforcement of hygiene standards, while TSOs focus on the protection of the consumer from unfair trading practices such as short-weighting and misleading labelling. In this chapter, we explore the changing patterns of food regulation at the national level, and this provides the changing backcloth against which local food regulators have been working in the 1990s. In particular, we draw attention to the emergence during the 1990s of a private-interest style of national regulation in the food system that contrasts with the traditional regulatory style based upon notions of the public interest. The emergence of a privatised style of regulation has been fostered by changes to the way in which EHOs undertake their work, for instance, by the promotion of the Hazard and Critical Control Point (HACCP) system.

Private-interest regulation

In order to try and understand the dynamics of the emergence of a private interest form of food regulation, it is useful to reflect upon the formative work on corporatism in the 1980s carried out by Streeck and Schmitter (1985) and their colleagues (such as Grant 1987). They were concerned to delineate an associative model of social order that was distinct from that of the community, the market and the state (Streeck and Schmitter 1985: ch. 1). An associative model is defined by its 'organisational concertation', and by this Streeck and Schmitter draw attention firstly to the functional organisation of group interests, and secondly to the integration of interest associations within the policy process. One of the leading commentators on corporatism at the time, Grant (1987: 1), noted that as work in the field progressed researchers became increasingly interested:

> in the question of whether business interest associations are capable of acting as private interest governments; that is, whether they are capable of discharging tasks that would otherwise have to be undertaken by the state, thus increasing the load of decision-making responsibilities which modern governments have to bear.

A private-interest model of government is of interest to the analysis of the contemporary regulation of food in Britain for a number of reasons. First, as we argue below, the British Retail Consortium (BRC) and its European counterpart, EuroCommerce, do play an influential role in the formation of food standards policy. Second, there is a hierarchy of retail interest associations with the BRC at the head. For Streeck and Schmitter (1985: 10), it is a central principle of the private-interest government model that 'concertation, or negotiation [occurs] within and among a limited and fixed set of interest organisations that mutually recognise each other's status and entitlements'. Significantly, and more problematically for our analysis, they continue: 'and are capable of reaching and implementing relatively stable compromises (*pacts*) in the pursuit of their interests.' As we shall see, it is not the interest associations, in this case the BRC, which deliver on implementation but the individual corporate retailers themselves. Third, following on from this, the private-interest government perspective does draw attention to the key role that private interests (i.e. corporate retailers) may play in determining and enforcing food standards within their own outlets and the subsidiary role of the public regulators.

For the state, a key advantage of moves towards private-interest government highlighted by Streeck and Schmitter (1985: 22–3) is that it helps to overcome the limits of legal regulation. An association that negotiates in the formation of policy (that is, their members' behaviour) has responsibility for its implementation and so a keen interest in ensuring that it can deliver on any agreements. Where a group is involved in regulating its own behaviour, it may also increase the legitimacy of the policy. Such an argument, of course, assumes that there is a public interest in regulating an activity, and as de Vroom (1985: 128) points out, additional factors could also come into play. These might include the complexity of the area and require considerable input of private knowledge, or pressures for deregulation. In the case of food, both issues are pertinent. Safety procedures and the composition of food products are becoming increasingly specialist areas with which public regulation and regulators have difficulty in keeping pace. Similarly, a commitment to deregulation was one of the hallmarks of Conservative governments (1979–97) and made an impact on the regulation of food (see Chapters 8–10 where this is explored in detail at the local level).

The work of de Vroom on the private-interest regulation of quality is of considerable interest. He applies his arguments to the pharmaceutical and food industries. Perhaps the most important part of his work is to distinguish the conditions for private regulation (a term he uses instead of government) and link them to the construction and maintenance of quality. It is worthwhile briefly following the outline of his argument and distinguishing where our perspective differs. He believes that private regulation depends on the presence of three features. The first is that the group has a 'degree of homogeneity and a *collective interest* in regulating quality' (de Vroom 1985: 129). In the case of British food regulation, however, it is not the BRC that will be involved in implementation but the individual corporate retailers. The critical issue here is that the corporate retailers do have a common interest in regulating quality through their own supply chains. But this common interest has to be implemented according to their own competitive retail-led supply chains. Competition is increasingly based around food quality rather than price. It is partly through their supply chains that the corporate retailers will be seeking to secure competitive advantage, and it is not therefore an area in which the BRC will be able to operate with any authority. As a consequence, retailers' perceptions of quality go above and beyond those of a state baseline standard.

The second feature that de Vroom identifies is the ability of an association to engage in quality regulation. This involves three elements: being able to bring members together, to be adequately resourced and to have some autonomy from the state and group members. In many ways the BRC does fit these features, but as we have argued above, in the case of British food regulation it is an asymmetrical relationship that exists between the corporate retailers, the BRC and the state: the BRC may play a part in policy making, but not in regulation. The BRC plays a significant role in negotiating with government on proposals that effect retailers and places considerable importance on being able to present a united face to government. It is able to act in the classic form of business interest association when its members allow it to do so, either through recognising a common issue or when they are prepared to make compromises to protect and promote their common interests. For any individual retailer, it may on occasions be a fine balance as to whether it believes its interests may be best represented by the BRC or through its own activities. As we showed in Chapter 3, the rise to economic prominence of the major retailers means that they themselves have the resources (e.g. money, knowledge and authority) to represent themselves to policy makers.

The third feature that de Vroom identifies is that the state must be in some way deficient in its regulatory strategy and must facilitate private-interest regulation. During the period of Conservative government, it was not only food but a broad swathe of government activity that was questioned. As we detail in Chapters 8 and 9, there were as a result noticeable shifts in the activities of local government food regulators. One of the consequences of the government's deregulatory strategy was that both through design and circumstance, the major retailers took on a more directive role in securing standards within their own outlets and through their supply chains for the products that they sold. In doing so, they were greatly assisted by the thrust of reforms to regulations which encouraged HACCP and the home authority principle, for example. The deregulating stance adopted by government – a clear admission perhaps of the deficiency of the modern state – came at a time of increasing corporate retail dominance. Both factors, rather than one or the other, came together to stimulate the conditions for private-interest regulation of food.

However, contrary to what de Vroom suggests (1985: 131), the development of more privatised food regulation has been directed towards procedures rather than products. The state has maintained its baseline standards and supervision, but over and above that quality standards are principally regulated and enforced by the corporate retailers. It is they who set standards for procedures and exercise control of their supply chains. What happens in practice, therefore, is that public- and private-interest regulation coexist, but they apply increasingly to different types of retailer. Below, we elucidate further the private- *and* public-interest regulation of food at the national level, and pursue this at the local level in Part III.

The private and public regulation of food

A central argument of James (1997) in his case for the creation of a centralised Food Standards Agency was that there were multiple agencies operating to regulate food with different reporting arrangements, styles and practices. His depiction of the regulatory apparatus is shown in Figure 7.1.

The James framework is a classic outline of the traditional understanding of food regulation. It is central government that sets standards through legislation, which are then enforced locally on behalf of the public by officials such as EHOs, TSOs and Public Analysts. They ensure similar baseline standards for all consumers. It is, therefore, a geographically bound form of

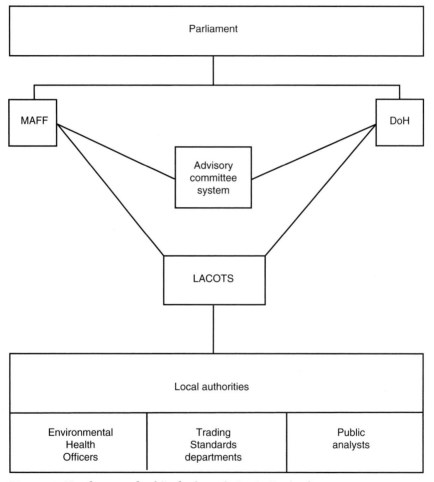

Figure 7.1 Key features of public food regulation in England

Source: based on James 1997: 12

regulation. In its more traditional and broadly defined form, food regulation has sought to ensure a combination of security of supply, accessibility, affordability and safety.

Crucially omitted by this model, however, is the impact that the major retailers have had on the regulation of food. James assumes a unified, relatively homogeneous process of regulation, when, in practice, as we shall see, a much more bifurcated system exists.

It is clear from the foregoing that there is more to regulation than its formal expression in statute or circular, and that it may involve more than simply governmental organisations. In the case of food, the ability of the major retailers to regulate their supply chains is certainly a significant factor in the operation of the food safety system. As James himself notes, the Food Safety Act 1990 not only consolidated much existing legislation but also implemented some European legislation: 'A key feature of the Act is the "due diligence defence" … In practical terms this means that food retailers have had to institute more extensive systems of checks on the foods that they sell' (James 1997: 11).

In practice, of course, the Food Safety Act simply put in legal form the supply chain regulation that the corporate retailers were already exercising, although for the non-corporate sector this would involve new practices (but not for all retailers, as James implies). It has also meant that due diligence applies unevenly. Regulators will undoubtedly have different expectations of appropriate food safety systems, increasingly exercised through HACCP, for different types of retailer.

The increasing differentiation between public and private sector regulation is illustrated below. As the model illustrates (see Table 7.1) the major food retailers, in contrast to the public-interest regulation undertaken by, for example EHOs, voluntarily regulate their own systems at their own expense, promoting individual choice based on their own hierarchy of quality definitions. So, as the major retailers have become the principal actors in the food system, they have negotiated key responsibilities in the management and

Table 7.1 A simplified model of public- or private-interest regulation

	Public-interest regulation	Private-interest regulation
Type of regulation	legislation guidance	voluntary
Form of regulation	central or local government	private or third party
Who pays?	taxpayer	manufacturer or retailer
Core values (in relation to the consumer)	(1) base-line recommendations	(1) individual consumer choice based on a hierarchy
	(2) obligations enforced on behalf of the public	(2) informed customers and quality definitions

policing of that system. Like all models, this public/private dichotomy is an abstraction, but it does highlight the changing nature of regulation and the role of the state and the corporate retailers in relation to food. Moreover, while we would not presume that these tendencies are necessarily replicated elsewhere, within the food sector, at least, they do illustrate well the process of state restructuring.

In the rest of this chapter, we elucidate the key role that food retailing and manufacturing and their representative organisations play in the construction of the model of private-interest regulation.

Regulatory change and the new 'battle for the consumer'

To appreciate the dynamism and the political and economic power that is central to the maintenance of the private-interest regulation of food, it is worth reflecting briefly on the changes that have taken place within the food system. From its high point some twenty to thirty years earlier (see Self and Storing 1962), the 1980s witnessed a gradual decline in agricultural producer corporatism in the UK. By the late 1980s, the monofocal character of the agricultural productivist policy was on the wane. Increasingly, it was the political distance portrayed between government and farmers, rather than the maintenance of a cosy producer corporatism, which typified farmer–regulatory state relationships (see Flynn *et al.* 1996). Paralleling these changes, and not unrelated to them, was the changing government stance with regard to diet and nutrition. Up until at least the late 1970s, the government had complemented its strategic priority of food production (increasing its ready and relatively cheap supply) with the postwar concern for improving the health of the nation via the encouragement of a balanced diet. This goal was seen to be best achieved through regulation, control and public education (see MAFF 1976; Darke 1977). The government took particular interest in vulnerable groups (Barnell *et al.* 1968) and recognised the direct role of government in promoting the benefits of a balanced diet. In short, it was very much part of the 'Fordist' regulatory system associated with food, with the industrialisation of agriculture and food being matched with a particularly regulated mode of food consumption which sustained male employment, gendered the household and created growing demands for household white goods (see Goodman and Redclift 1991; Fine *et al.* 1996). By the early 1990s, it was clear that the competing interests involved in the food sector had broadened, diversified and reshaped themselves round the three leading features that we have developed.

At this time, the Conservative government was trying to find a more limited definition of the role of the state, corporate retailers (as we have documented in Chapter 3) were clearly outstripping food manufacturers in their rate of growth and return on capital, and many of the postwar problems of supply and scarcity had been seen to be solved. Under these conditions of more limited state intervention and control over the supply of food, there grew a new phase in the regulation of food, one which was to become much more complex from a regulatory point of view, and one which was increasingly having to acknowledge the spreading of food quality consciousness.

The backcloth against which the representation of retailers and manufacturers – the key representatives in the contemporary food system – occurs was formed first by the growing Europeanisation of food policy (see Chapter 2); second, by the emergence of food issues from agricultural and more recently single market policy; third, by the lack of government commitment to more rigorous regulation in the public interest; and fourth, by the gradual and secular growth in the public consciousness about the quality and types of foods from which consumers were able to select (the social provision of choice). This has led to the emergence of a more complex policy community of food in Britain.

Today, the representation of the consumer interest is no longer simply the responsibility of government, or of government working with particular sectoral interests like farmers (see Self and Storing 1962) or the agricultural research and development establishment. Representation and the articulation of the consumer interest is much more complex and reliant upon a matrix of competing interests which, although we have portrayed them in ideal-typical form as a neat dichotomy, in practice do not fit neatly into public or private sectoral divisions. We will now deal specifically with describing the key retailing and manufacturing parts of this community. As we have already argued in Chapters 5–6, we are witnessing a situation where consuming interests are taken up and articulated in relation to the supply chain itself, and this tends to give particular opportunities to the ascendant corporate retailers who, positioned as they are, at the sharp end of point of sale delivery can begin to represent the consumer in ways which enhance the retailers' role in regulating the supply chain itself. The representation of the consumer is thus the outcome of two general and somewhat contradictory trends. These are (a) the evolution of market-based food supply systems giving particular powers and opportunities to the near-market agencies (particularly the corporate retailers), and (b) the growing public concern for food quality as the series of food crises have been experienced. As a result, we now have a food system which is much more conditioned by competing interests and competition; not only in the attainment of markets, but also in the battle to attract and legitimate consumer loyalties. The food policy community is therefore more driven by the varying needs to articulate, represent, study and manage these risks and the knowledge on which they are based. Much of this knowledge is based around the construction of the heterogeneous consumer in mind (although Warde (1997) cautions against exaggerating here).

We explore below how key economic organisations involved in the food system seek to convey a legitimate conception of the consumer to government, and try to be seen as the legitimate articulators of their interests. This in turn provides a key support for the regulatory activities that occur within the private-interest model. Here then we have a major dynamic through which the political and social action of the food system operates, and a basis for marking out the contested terrain in which retailers in particular, but also manufacturers, operate.

Retailers

For the corporate retailers, collective representation has been a somewhat more reluctant and recent process than that for food manufacturers. Intense competition between themselves, and with manufacturers and other agents in the food supply system, has somewhat naturally forced them to 'look after themselves' as far as representation of their interests is concerned. However, the growth of careful consumption in the 1990s, and by dint of their expansive role in social and economic restructuring in Britain, meant that retailers were increasingly confronted with new sets of issues common to all the major corporates. On some issues (such as Sunday trading, deregulation policy, labour law, or law and order) retailers could collaborate, so as to present an even stronger case to the government based on their economic success and social power in the consumption sphere.

The BRC and its European counterpart – EuroCommerce – have developed as important lobbying agencies for the major retailers (see Marsden and Wrigley 1996). In the 1990s, these relatively new trade associations began to provide a valuable coordination role for individual retailers at a time when they could potentially come under much pressure from government and the public. Their role as consultees in the national government and EU policy debate became much more significant. They have developed close links with officials at both the national and European level. Food scares and increasing threats from competition policy and food safety legislation (see Wrigley 1995; Marsden and Wrigley 1995) have made it necessary for the retailers to have a permanent and more coordinated lobbying force in order to put their case to ministers and civil servants. In particular, as we document later in Part III, retailers have been keen to ensure that they are regulated in a consistent manner. Hence, the food and drink committee of the BRC was particularly active. During 1993–4, for example, its priorities were seen to be the deregulation of food law, implementation of the units of measurement directive, food hygiene issues, especially the implementation of the EU Food Hygiene Directive (Council Directive 93/43/EEC) and the amendment of British legislation governing food temperature controls, the effectiveness of the UK food retail representation in the EU through EuroCommerce, and laying the foundations for BRC to influence discussions at the EU Agricultural Council level. All this involved articulation of the 'real world' of corporate retailing to government, to ensure that it understood retailing, its complexity and vulnerability:

> The committee's initiative to provide MAFF officials with experience of how food retailers operate was judged a great success, and a further round of visits was organised with more retailers than before agreeing to participate. BRC understands that there is a waiting list of MAFF officials building up for future visits, which last for three to five days depending on the size of the company.
>
> (BRC n.d.: 1)

It is not just that the BRC can bring an understanding of retailing to government, it is also centrally involved in devising means to implement new initiatives:

> BRC submitted views to the DoH [Department of Health] on draft regulations to implement the Hygiene Directive which DoH hopes to bring into force a year before the due date. BRC has also set up a new working party to draw up the Guidelines for Good Hygiene Practice for Food Retail under the provisions of the Hygiene Directive. The working party will be working closely with DoH, LACOTS [Local Authorities Coordinating Body on Food and Trading Standards] to ensure that the Guide meets all the requirements for maintenance of a high standard of hygiene in food retail businesses of all sizes.
>
> (BRC n.d.: 1)

BRC is therefore engaged in constructing a web of relationships not just with government, but with other actors in the food system to protect and promote the interests of retailers. For example, in relation to LACOTS it has claimed:

> BRC enjoys an excellent working relationship with LACOTS and there have been distinct improvements in the way in which food law has been enforced … There is greater emphasis on informal guidance than on legal action as a means of achieving better compliance with the law. BRC has welcomed a decision by LACOTS to consult interested parties on controversial issues on which LACOTS intends to issue guidance notes for enforcement authorities on how to interpret certain legislation.
>
> (BRC n.d.: 1)

Similarly, the BRC has close contacts with the Food and Drink Federation and will often seek to negotiate agreements of mutual benefit rather than have a decision imposed by government. The following extracts from an interview with a BRC representative illustrate the emerging legitimacy of BRC with government and with the retailers themselves:

> I think retail and trade organisations are always important. But, it's a question of whether they are recognised as being important by their

members, or their potential members. Certainly in the European Commission and government departments, they like to deal with trades associations because they like to get, in so far as possible, a general view of proposals for legislation etc. that they can say is representative of the industry as a whole, rather than one particular company's view. The latter's view might be biased because that particular company might be extremely good at something and it might want to get the legislation biased to help them, knowing that others will have difficulty in complying with it. So, in that sense, we have a lot of power because the government departments and the European Commission like to hear from an umbrella organisation that is able to show that it is representative of that group of retailers, or whatever it may be. We have got a very comprehensive membership. We represent 90 per cent of retailing by volume sales. That's big. The area where we are weakest is the independent retailers. There are a large number of independent retailers who don't belong to trade associations because they don't understand the importance of them. It's either that or they don't want to pay the subscriptions or whatever. So they are not plugged in to what is going on as a result. But their share of the market is so small that as far as we're concerned, although we like to represent the small trader's view as well, and to bear it in mind when formulating our policy, they are represented by other trades organisations as well.

I mean government needs to be able to legislate in a way that is practical for the industry to comply with, so they obviously rely on the industries concerned to inform them if they think that what is being proposed is not feasible. Obviously, they welcome that sort of input. They feel sometimes that they need to read between the lines because a lot of trade organisations exaggerate in order to make the point. I try very hard not to do that because I think it is essential to maintain credibility with the government. They can tell if you are exaggerating because they always investigate who is making a point, and they find out who is exaggerating, and the next time, they are less likely to find that person credible. So I'm always telling members, 'if you don't like something, say why you don't, and provide accurate facts and figures to back up your point'. You see, sometimes you can't give very explicit reasons why you don't approve of something, but if you fail to give good reasons, then you find that you don't get very far.

Interviewer: Going back to the issue of the consumer again, you say that one of the reasons why government listens to you is because, in a sense they see you as representing the consumer's view indirectly. How do you know that?

Well, because the large retailers, in particular, have day to day contact with their customers, and if their customers write to them and say what they do want and don't like, then all this information is kept by the retailers, so that they have a very good sense of what the consumers want, and also they get a lot of information from the sales. They may, for example, get people saying, 'well, you know, why don't you take sugar out of everything?' – I mean I'm exaggerating here to make a point. Well, the retailers know very well that the final product would not sell. The retailers know what the customers like to buy and what sells. So, clearly that information is helpful. I mean there are all these do-good organisations who are trying to influence the health of the nation.

Food manufacturers

The main food manufacturers group is the Food and Drink Federation (FDF). It has a specific remit: 'to encourage consistent attainment of the highest practicable standards by everyone in the food chain from the farmer, via manufacturer and distributor, to the consumer ... to be involved actively in the shaping of future legislation in the EU and the UK and to participate fully in its development and implementation' (Food and Drink Federation n.d.: 44).

It is a well-resourced organisation and, for example, was at one stage fielding twenty of its industry representatives on key project teams set up by the Department of Health's Nutrition Task force, established as part of the *Health of the Nation* White Paper. It had also inaugurated a 'lifestyle' campaign, with panels of leading scientists and nutritionists to 'provide continued support in the development of implementation of the campaign; and to build alliances and partnerships with health professionals, schools, the sports and leisure industry, local authorities, the media etc.' (Food and Drink Federation n.d.: 44).

Accepting that consumers need to become fitter not fatter, the FDF claimed the consumer was their main concern. Fighting the battle for the consumer on a variety of fronts is however required. In particular, promoting the potential value, as opposed to the contradictions inherent in new technological innovations, has become a major function. The organisation seeks to persuade and legitimate both consumers at large and the plethora of other, less sanguine organisations in particular. This is a contest over the legitimacy of knowledges and rights about foods. The FDF, representing food manufacturers, engages in shaping public perceptions.

In 1995, against growing concern over biotechnologies and genetic modification, and with Sainsbury being the first to test out the reaction by putting genetically modified tomato paste on its shelves, the FDF initiated the 'Public Perception of Biotechnology Working Group'. This body held prestigious seminars, conferences and panels on issues of food and biotechnology, creating 'positive responses from many of the organisations and individuals with whom FDF was not a regular contact, including representatives from religious, ethical and environmental groups'. This attempt to reach out to the potential opposition, and the media management that is associated with it, is now a key feature of the main economic-based interest groups. For these groups, food representation is big business, and it has increasingly to reach new areas of interest mediation both in Brussels and at home. This is now the meat and drink of competitive market maintenance in conditions where careful consumption by the public is matched by a hungry media keen to expose the intricacies and risks associated with food consumption. The mode of articulation involves sophisticated conceptions of what is in the best interests of the 'consumer'. These organisations have to continue to achieve at least some legitimacy in these battles over the consumer concern.

The British Nutrition Foundation

Allied to the determination of different interests to portray a legitimate representation of 'the consumer' is a concern to address and position the respective organisations in relation to both existing products which may come under scrutiny, and to new innovations emanating from retailers and manufacturers. A particularly good example is the British Nutrition Foundation (BNF), which is funded by over forty food manufacturers and retailers. It prides itself on:

> being an impartial scientific organisation which provides reliable infor-
> mation and soundly based advice on the nutrition and related health
> matters. Its principle objectives fall under the headings of information,
> education and research ... its activities and services are extensively used
> by the health care professionals, consumers, industry, educators, journal-
> ists, government departments, the media in all its forms.
>
> (BNF 1995)

The BNF has particularly found the need, which is not unrelated to those of its sponsors, to articulate the advances of biotechnology and genetic engineering to 'the consumer', and more particularly to the consumer and media lobbies. It promotes itself as a forum for the legitimate base of food science to get its point across to 'non-scientists'. Underlying this has been a strong emphasis on the legitimacy of science over public anxiety about foods and innovations in foods. The BNF represents an attempt to introduce 'order' into the food world, which is increasingly subject to much closer public scrutiny and scientific innovation. In addition, it has to accept the highly competitive market delivery of food goods as a given.

Reflections on regulatory change

While for the representatives of the food manufacturers, such as the FDF, the main objective has been to balance the interests of consumers with attempts to gain acceptance of new innovations in food manufacturing and supply, for the retailing organisations, the articulation problematic has been much more diffuse and variable. This depends on influencing the evolution of food quality policy as it affects consumers at the point of purchase. This was increasingly necessary for the retailers because this locus represents their main cash nexus for accumulation: that is, the point of sale. This again is in stark contrast to the postwar command and control days, as retailers are now subject to potentially high degrees of volatility and competition. It is thus more critical for the retailers to create room for competitive manoeuvre; that is, to develop their competitive space through the legitimate and continual construction and representation of the consumer interest.

Increasingly, given the growing raft of EU legislation as well as the implications of the internal market for retailing, EuroCommerce is beginning to play a much more functionally important role. The following statement from a recent EuroCommerce document sums up the ways in which the retailer organisations construct their own internal notions of the consumer interest, and why in the 1990s they became so involved in the regulation and representation of consuming interests. It also demonstrates the considerable distance travelled from the assumptions of the Fordist mode of collective food consumption associated with macro social provision in the early postwar era.

> Since the early 1990s retailers have had to deal with an extremely differentiated consumer profile. The typologies of the 1970s and 1980s, which were based on the socio-demographic characteristics and lifestyle forms, are no longer able to adequately describe consumers' apparently volatile behaviour. In all European countries, opinions, attitudes and the consumer behaviour associated with them are increasingly shaped by:
>
> 1   a surfeit of media and global information about the accelerating changes;
> 2   growing awareness and concern about global problems and the local impact of consumption on the environment, human and animal health, social relationships, and jobs as well as a growing willingness to take appropriate action;
> 3   professionalisation in certain fields of consumption (e.g. hobby, cooking, do it yourself, home gardening, sports) has differentiated and specialised demand for a wide and complete range of the corresponding products and services; and
> 4   a conscious, case by case choice between expenditures on luxury goods, value for money and the cheapest offer. This has become more pronounced in Europe since the disposable income of wide sections of the population has started to stagnate.

To retain customers' trust and loyalty, retailers need to provide answers and solutions for customers' varied demands. This in turn requires institutionalised forms of dialogue and ongoing processes of innovation in product range, formats, communications and services that go far beyond traditional forms of market research and product advertising. In today's world of abundant choice, consumers select the shopping location that offers them the best solution to their problems and the most pleasant shopping experience. *This is the new dimension of competition in quality and image that retailers have to confront* [our emphasis].

(EuroCommerce 1997)

Rather ironically, despite all the 'improvements' in the efficiencies and retailer logistics, and despite the orchestrated rationality of attempting to meet consumer preferences, corporate retailing in the 1990s is forced to see consumers, and to project images of them, as much more socially and politically constructed and influenced. As a result, any rational attempt to continue to gain profit, to accumulate from the retailing process, needs to engage deeply with the social and regulatory institutions which are involved in food consumption, for they provide the terrain upon which the maintenance of 'retailers' competitive space' is fought. Such is the complexity, the volatility and the public consciousness today of food issues that a more straightforward reliance on corporatist governance, equivalent to that experienced by farmer groups in the postwar period, can no longer suffice. It is the public and private dimensions of regulation that we must now seek to understand. Government, having divested itself of a central and guiding role in coherent food policy making, is now faced with the tentative realisation that any crises of legitimacy which may occur can only partly be rectified by the state itself. Food regulation, while much more complex, is also highly prone to legitimatory crisis and dysfunction. Corporate retailers have become a major agent in fending off such crises. The construction of the consumer interest is now central to the dynamism of the food policy community, but these constructions are privately as well as publicly articulated and used in the making and maintaining of corporate retailers' competitive space.

Conclusions: the co-evolution of regulation

As we argued in the previous chapter, British retailing is highly competitive and heterogeneous, and this has important implications for regulation practices. At the peak are the 'big three', Tesco, Sainsbury's and Safeway; the second tier includes Gateway/Somerfield and ASDA; the third tier are discounters like KwikSave, Aldi and Netto; and the fourth tier includes the more numerous but small and declining independents. Associated with these different tiers of retailing are different constructions and regulatory practices concerning food quality. It is the corporate retailers that have led the way, not only in terms of innovative forms of competition and food design, but also how to regulate food quality under increasingly complex food supply chain relationships.

The food hygiene and hazard systems operated by the corporate retailers have increasingly become a condition of market entry for food suppliers and manufacturers. Hence, as far as the overall supply chains are concerned, it is no longer enough from the point of view of the retailers to supply quality food of the right compositional standards. It is also necessary for suppliers to demonstrate that systems of quality management have been put in place (such as the Hazard Analysis and Critical Control Point system (HACCP)) as a food assurance scheme. Hence the retailers expect more and more from their suppliers in terms of the policing of food delivery as well as the type specifications of the food produced. This stands to give retailers a market advantage with customers, and it demonstrates to government that they are taking existing food regulation seriously (particularly the 1990 Food Safety Act and the 1996 Food Hygiene Directive). For the customer, this means that food quality is both highly differentiated above the state-defined base-lines, and that the choice of retail outlet will affect the type and the constructions of quality purchased and consumed.

A key element in the ability of the major retailers to engage in self-regulation is that they work in tandem with public regulatory activities. This is shown in Figure 7.2. Three innovations in the 1990s have helped smooth the process between public and private styles of regulation, and these are explored in greater depth through their implementation of food policy at the local level in Part III. One of these is the *home authority principle* (see Chapter 9). The other, and subsequent, new procedures and techniques of assessment were *industry codes of practice* and *hazard analysis*, and both provide further legitimation for the hegemony of retailers' supply chain management. The promotion of both codes of practice and of hazard analysis allows EHOs to differentiate, in their regulatory activity, between the 'superleague' and other food retailers. Crucially, it represents an acknowledgement by government, and a willingness of the 'superleague' of retailers themselves, in large part to engage in self-regulation. In other words, it is a key step in retailer-led moves towards private-interest regulation. In practice, as we shall see in Part III, EHOs now increasingly adopt a bifurcated approach to the regulation of food retailing. They largely retain their traditional approach, mixed with some rudimentary hazard analysis, to the bulk of independent traders. For the major retailers, however, EHOs are adopting an auditing approach in which the supermarkets' management systems are tested.

The outline of such models here also has important implications for our understanding of private-interest regulation. It is important to distinguish between private-interest *government* and regulation in two ways. First, the former is a much broader concept and has more significant implications for the structure of governance in that it is concerned with the ways in which business interest associations may contribute to both the making and the implementation of policy. Private-interest *regulation*, as we have used it here, is much more concerned with the ways in which policy is implemented. We recognise that the BRC can be a significant player, under certain conditions,

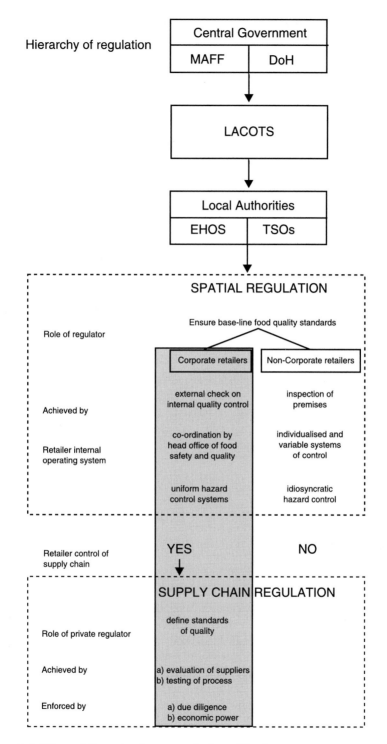

Figure 7.2 The public and private regulation of food

in dealing with government, but that the corporate retailers are also well able to present themselves to, and be listened to by, government. Second, a feature of private-interest *government* is that business interest associations take on roles and responsibilities that governments would otherwise have to perform. As we have shown in relation to food, the situation is now somewhat more complex. The corporate retailers were not assuming a role that government would otherwise have undertaken, because government continued to play an active role. Rather (as we saw in Chapter 6), by going beyond the state baseline, the major retailers have been able to formulate and impose their own food quality standards. In doing so, they can argue that they have performed a public interest role and one which may have at least partially relieved government of its regulatory burden. This is because the major retailers are such dominant locations for food purchasing that their activities concerning food standards have highly significant implications for the quality of food that is purchased. Thus by pursuing their intensely competitive interest in market maintenance, the major retailers have also performed a public-interest role in relation to food standards. They have also for the most part proved much more able to assuage public fears over the safety of food than government. By doing so, the corporate retailers may for a period of time have lessened the need for, and extent of, government intervention.

Thus far, our study of the interactions between food policy making and its implementation in the 1990s has been pitched at the national and European levels. In the following chapters, we analyse the process by which the nature of local level regulation has been redefined to accommodate these macro changes in the nature of retailing and the government's deregulatory stance. In this way, we describe and explain the introduction of new techniques of food regulation at the local level that have eventually consolidated a more dichotomous system of food retailing, which allows non-corporate retailers to be publicly regulated and the 'superleague' to regulate much of their own operations with minimal external checks. In short, we explore the different spheres of spatial and supply chain regulation (as indicated in Figure 7.2) to meet Clark's (1992: 616) call for 'finer-grained research' of regulatory practice.

As we shall see, a significant outcome of emergent regulatory practices has been to show that major retailers have been keen to gain uniform treatment of their outlets from the multitude of geographically variable local authorities in which they are located. They have been largely able to achieve this through the adoption of the home authority principle and the application of hazard analysis. Moreover, the ability of such retailers to manage the quality of foods through their supply chains has also been enhanced by the passage of new regulations and as the growing crisis of confidence in food consumption more generally has taken hold. As a result, new relationships have been forged between public and private regulation. It is increasingly apparent that a public system of regulation that has defined geographical (local authority) boundaries is increasingly irrelevant to the diverse sourcing strategies and nationally regulated regulatory abilities of the major retailers and the foods that they sell nationally. For the large number of smaller, independent retail outlets the situation is, however, quite different. They too must now make some efforts to identify and control hazards within their operations, although they are likely to do so on a more traditional basis. It is here that state regulation remains important for maintaining food standards. It is the co-evolution of these public and private regulatory systems that are represented in Figure 7.2, and we outline them in much closer and local detail in the next three chapters of Part III.

Part III
Local strategies

8 Local retail-consumption spaces and hierarchies

Introduction: contested retailing geographies and food provision spaces

The previous parts of the book have documented the emergence over the last two decades of corporate retailers, who have come to dominate food provision in the UK, and at the same time have developed a new embedded relationship with the national state. In this part, we now turn our attention to the local articulation of these developments, exploring the manner in which food provision within individual retail outlets is regulated by public and private interests, and the consequent relationship between retailers and the local state.

The analysis focuses on how the retailer–state relationships outlined earlier are translated and implemented at a local level; or, in short, how such new regulatory relationships are 'spaced'. This chapter serves both to introduce the micro focus and to expand upon the macro theme. In depicting the local retail geographies of the study areas, it introduces the actual environments within which the analysis of micro-regulation will be based. And in so doing, it also explores the way in which national and international changes associated with the corporate retailers have impacted locally. The discussion, then, is concerned with depicting the recent history of change and contemporary structure of food retailing, as described in Part I, and as it is witnessed in each of the study sites; and also with the local formal organisation of food regulation in these food retail spaces. The chapter begins with the exploration of notions of 'regulatory space' that are expanded upon and clarified through the subsequent Chapters 9 and 10.

Much of retail geography has largely involved empirical work which describes analytically the changing spatial structure of retailing (Clarke 1996: 286; Wrigley and Lowe 1996). In this chapter, however, we are interested in describing the changing structure of retail provision in the field sites but, more importantly, we are concerned with the nature of the relationship between the retailer and the local state, and the latter's relative ability to influence the local geography of corporate restructuring. As Clarke (1996: 287) notes, the shift from 1979 onwards towards less 'restrictive' retail planning effectively meant that the central state withdrew from urban planners the right to regard their paternalistic judgements of equitable retail provision as superior to the outcome of the market. This change has certainly not stemmed the battles between the retailers, planners, residents and other interest groups, but there can be little doubt that since the late 1970s retail planners have had the merest vestiges of their former powers. Retailers have long since learned to play the system, and planners, often out of local economic necessity, are often prone to give way to 'job creating' or 'infra-structure-improving' retail developments. These promises have themselves, as the discussion elaborates, created new debate regarding the local roles of retailers. So, amidst these reconfigurations of a 'new' retail geography, we introduce here the patterns of change in local food provision in the study areas, and consider the factors that have affected its timing and outcome in different places.

The second section of this chapter provides a brief introduction to the field sites of Newton and Stonefordshire, describing their location, popula-tion and broad socio-economic characteristics. The third section explores their recent history of retail change, its outcome and the contemporary patterns of food provision. Using evidence from interviews with local plan-ners (i.e. those public officials responsible for siting the stores), it considers the local state's attempt to, at best, harness the potential of corporate power for local economic development; and at worst, ameliorate its negative impact upon non-corporate retailers and vulnerable sectors of the local population. The fourth section introduces the local formal organisation of food regula-tion: the functions and responsibilities of those professional groups – environmental health officers, trading standards officers and public analysts – involved with the local level enforcement of food legislation, and the food premises within the food authorities' jurisdiction.

The study areas: local and contested regulatory spaces

Food retailing is regulated within local food authorities. City-based *unitary* authorities operate at the borough level along with all other local state functions. The *non-unitary* authority structure of the shire counties, conversely, includes a lower tier of district authorities which operate under the county council. In this case the food authority functions are undertaken at the district level. The two study areas here encompass both circumstances. The first is the London borough of Newton, where EHOs and TSOs work within a concentrated inner-city retail environment of a unitary authority. The second is in Stonefordshire, where a sub-tier of six district authorities functions under the county council. The study involved four of these latter, two of which were city districts and two of which included retailers located in smaller towns and villages.

The borough of Newton was formed in 1965 from the old metropolitan boroughs of Wellford and **Newton**. Apart from its numerous parks and recreation sites, it is an entirely urban area: 230,983 residents live within 13.7 square miles. Within the borough, however, as is typical of London, there are fourteen 'villages' and towns that each have their own distinctive characters and history. Newton itself is the main focal point of the borough's economic activity. It is ranked fifth in the list of London's district centres, and is the only centre of major regional significance in inner London.

Newton borough is a multi-cultural community: one in five of the population are from black or ethnic minority communities. It is also an area with a high incidence of social and economic deprivation. Indeed, it was ranked twelfth among London boroughs in terms of multiple deprivation in the 1981 census; average household income of residents in 1985 was below the national average and only two-thirds of that of the Southeast region; and while 40 per cent of households in 1987 received housing benefit, nearly one-third were receiving income supplements (Hyde *et al.* 1989). The borough also has higher proportions of unskilled, semi-skilled and skilled manual workers than London averages.

Historically, a division has arisen between the relatively poor area in the north of the borough and the more affluent areas in the south. Unemployment is highest in the north, at 24 per cent compared to 17 per cent in the borough as a whole. The north also has the greatest proportion of council tenants, residents from social class D and E and one-parent families. Thirty-eight per cent of the population of this area are from ethnic minority groups (Keane and Willets 1995). These are most concentrated in Wellford, which is one of the borough's most densely populated areas and has the highest unemployment. In an effort to tackle some of these problems, this area was awarded a City Challenge grant in 1990. However, the picture is not entirely one of deprivation: Newton includes many prosperous districts and income levels are rising steadily among those in work; and the population of the catchment area for Newton town centre, which stretches east into Harbourville and Brambley, includes many high-income households.

The county of Stonefordshire, by contrast, is centrally placed in the heartland of 'middle' England, covering an area of 1,025 square miles. The quality of the county's landscape is reflected by the designation of large parts of it as Areas of Outstanding Natural Beauty. In this glorious setting, many historic villages and small towns are found.

Roughly half of the county's 552,731 residents are urban: the county capital, the city of Stoneford, has a population of 105,822 and the town of Cettleham has 106,575. The county's other major towns of Shington and Tetington have 23,336 and 9,595 respectively. The local district authorities cross this divide: Cettleham and Stoneford authorities are entirely urban, Shington and Tetington include urban and rural areas.

While Stonefordshire is relatively prosperous, with concentrated areas of considerable wealth, the socioeconomic status of its population varies across the county. Approximately two-fifths of its population as a whole are in social class A and B, with one-sixth in categories D and E. In the visually charming towns of Cettleham and Shington, nearly 50 per cent of the population are in the former group; but in Stoneford this falls to less than one-third. The district of Tetington, with its splendid medieval town and appealing farmhouses, also has a high proportion of middle- and upper-class residents. Much of the county's poverty is rural.

The rise of the corporate retailers in the UK and the concurrent decline of independent shops has undoubtedly produced a certain uniformity in food provision nationally. It may be expected that even minor towns have acquired at least one 'superleague' food retailer over the past fifteen years (likely to be located on the edge or out of town), and concurrently, have witnessed a decline in high-street butchers, bakers and fishmongers. For instance, nationally from 1985, more meat products were sold out of supermarkets than from independent butchers. However, it is perhaps remarkable that it is quantitatively impossible to document these changes exactly in most local authorities. Local surveys of retailing premises undertaken by the planning departments in the study areas were abandoned in the 1980s as a result of 'financial restraints' in local authority funding. Some private companies engaged in market research have business data based on sample surveys, but this is generally unaffordable for public sector organisations. In short, the local authority planning departments involved in the study areas are unable to document the exact change in their local retail geography. While it is easy to count the number of new 'superstores' and new retail developments, which have had to seek planning permission, it is impossible to portray the precise nature of change amongst the 'lowest' tier of retail outlets. The planners involved in the study all believed that they had a very good 'feel' for the change that had taken place in their areas, and they also had 'proxy' indicators, such as employment statistics in the retailing sector to use. These assumptions are based, however, on poor local data of the rate and type of retail change. Corporate retailers tend, on the other hand, to undertake extensive catchment analysis before considering development.

Our interest in the nature of food retailing in the study sites is twofold: we are interested in the manner in which the structural changes that have taken place in food retailing at the national level have impacted locally, and we are concerned with the contemporary nature of the food retailing that local authority food law enforcement officials are responsible for regulating. In the following discussion we therefore utilise two different types of data. Below, we explore retail change and the relationship between the local state and corporate retailers through the eyes of the local authority planners themselves, using extracts from the transcripts of our interviews. Later, we use data supplied from local authority food regulators to look at the contemporary nature of retailing in the study sites.

Since the 1990 Food Safety Act, businesses involved with the sale of food have been legally obliged to register with the local council. Environmental health and trading standards departments keep computer records of this, which are able to identify different types of retail outlet. Using this data, we are able to provide some indication of the contemporary configurations of food retailing in the study sites. Through this analysis, we can attempt to gain a picture of the character of retailing hierarchies in the two study sites.

Food retailing space

The two field sites of Newton and Stonefordshire comprise seven food authorities: two trading standards departments (for Newton and Stonefordshire), an environmental health department for Newton; and four environmental health offices at the borough level for Shington, Tetington, Stoneford and Cettleham. The figures given here are from the records held by these offices regarding the number and type of food premises under their jurisdiction. They illustrate the variation in the regulatory environment of the different authorities, and by default, the relative 'tiering' of food retailing.

Table 8.1 Retail hierarchies as recorded by local enforcement officials

Stonefordshire trading standards (county wide)	Stoneford environmental health (county town)	Newton trading standards (London borough)	Newton environmental health (London borough)	Shington environmental health (county town)
3,457 registered premises comprising:	911 food businesses comprising:	1,264 food businesses comprising:	2,011 food premises comprising:	1,314 food premises comprising:
77 super-markets	7 supermarkets	77 super-markets	78 super-markets	19 supermarkets
509 grocers	254 retailers	570 restaurants, caterers or takeaways	129 grocers	39 butchers
129 fruit & veg shops	17 manu-facturers	808 butchers, bakers, fishmongers, green grocers etc.	47 butchers	19 greengrocers
120 butchers	1 packer		52 bakers	152 gen stores
91 bakers	32 distributors		All the rest takeaways, caterers, school kitchens, cafes, restaurants	883 caterers, include public houses, hotels, etc.
	607 restaurants and caterers			65 manu-facturers

Retail provision and local planning: conflict and coalition

Newton: 'bucking the trends' in the inner city

Food retailing in the borough of Newton is based around a shopping hierarchy consisting of the major 'strategic' centre of Newton town centre, and a
number of district centres, some smaller local centres and local parades and
corner shops. Retail change and rationalisation began very early in London.
Although the town centre is still the most important location for shopping,
the borough has experienced change from the mid-1970s, with three consec-
utive 'waves' of shopping decentralisation. The first wave was associated with
the introduction of free-standing superstores and hypermarkets. The second,
starting in the early 1980s, was associated with the growth in free-standing
retail warehouses and more recently, the grouping of such stores into 'retail
parks'. The third wave of out-of-centre retail development dates from the
mid-1980s and is associated with the large numbers of proposals for comprehensive retail shopping centres (Unitary Development Plan 1993: 284).

There has been long-term decline in the number of shop units and the
concentration of remaining facilities in the larger centres. Local centres have
grown organically with the surrounding neighbourhoods, and some are now
in decline; that decline being most severe in those shops selling food.
During the 1970s, when the borough was still undertaking retail surveys,
and between 1971 and 1976, the number of shops in the independent sector
was shown to have declined by 39 per cent. However, this has not been
entirely attributed to the impact of bulk buying in supermarkets. Broader
socioeconomic factors are also judged to be important: the decline in the
borough's population (typical of inner-city London) has removed spending
power, and demolition as part of wider redevelopment schemes has also
affected shops. A local planner describes these phases and their impact upon
food retail structure in the borough:

> as far as closure on the high street and the expansion of big stores ...
> well, it's no different really here to elsewhere. Our surveys get less and
> less frequent, so it's hard to be exact in terms of numbers ... We used to
> do a comprehensive land use survey. Every building in the Borough was
> noted and its use noted and then we were able to say fairly categorically
> what was going on, and it really goes back to the mid to late 70s, and
> we know that a third of all the grocery outlets in the borough are now
> closed ... these are mainly High Street ... or corner shops. The high
> street decline also coincides with the phase of rationalisation of the
> multiples, so if there was a small Sainsbury's, that will have gone and
> also Tesco, as they went out of the centre. We look at employment data
> in the retail sector; we know that between '81 and '91', that also went
> down. It's not obviously an exact surrogate, because of efficiency levels –
> but presumably it's some sort of guide. So that's the picture that we

have ... I think that what there has been is closure of the small firm stores, the major multiples getting larger and larger and the rationalisation of some of the small firms as well. There's also been in more recent years a segmenting of the market, so that having gone hell for leather out of town, with the big stores, 60,000, and 35,000 sq. feet sales area, and closing the smaller in-town ones, the kind of budget or cheaper – who initially were KwikSave and Iceland – came in to town centre and filled some of those spaces.

In such a deprived urban area, this has further marginalised the most vulnerable:

> We have a slightly older population, particularly in the Mellville area, so it's geographically different so if you take Mellville for example, those older people might suffer because they like to use the more local shops ... they may have suffered for the very local shops closing ... If you take Wellford, which has got the higher concentration of younger people and Afro Caribbean, they have lost their supermarket, so they are probably travelling to Newton or to Surrey Keys – you can see when you visit these places in towns, you can see – that people without use of a car are queuing up for the minicabs, so it's presumably added to their expenses. So Newton's policy then, is very much formulated around the basis that it is a borough in which people have got low socioeconomic standards and don't have access to their own transport.

Related to this is the planners' aim to maintain the existing shopping hierarchy and reinforce Newton town centre as a strategic point. Indeed, to this end a Town Centre Action Area Plan has been a major concern for the council. This has involved the provision of resources to 'realise and consolidate its status', and to counteract its lessening importance in retail provision. It is a reactive and palliative strategy to the retailer-led spatial restructuring and operated on behalf of the 'disadvantaged' shopper; but as the planner explained, it *ameliorates* the consequences of retail development rather than directs it:

> our policy is because the town centre is a transport nodal point, we argue that that is an equality issue, that more people can get to those centres, that's where the stores should be located. The big chains argue that it's all about the car, and from their point of view it is – because that's where the money comes from ... so we adopted policies in our Unitary Development Plan to locate stores in or adjacent to town centres. And we did fight appeals with the multiples, on their efforts to go out of centre. Some of which we win, some of which we lose ... *But you know, there are sticking points, we have to believe this don't we ... that our planning policy helps to shape what would happen anyway. I mean, you know,*

it's very difficult for us to really buck trends but we shift around at the margins. So we designate what we call the primary and secondary areas in the High Street, and said, you know, you've really got to persuade us to change our plans for shop use in the core area, but we'll be more flexible in the secondary zone. Instead of having 20 per cent vacancies dotted throughout the thing, you try to concentrate those so it'll look like a viable shopping centre. So that's how we responded through the 80s [emphasis added].

In this respect, the local authority will try to enlist the help of the retailers in achieving their aims:

Certainly, if they (the top-tier retailers) want to develop in town, then we do everything we can, you know. Money makes it difficult now. In the past it would have involved compulsory purchase, things like that ... because that's what we want. They are the transport nodes where everyone can get to. We argue that that's the fairest way. Only 50 per cent of people have got cars – so people can go in and do a bit of shopping. We'd like to see those centres reinforced, and we'll encourage that. If it's not we'll fight it as much as we can, or if it still seems reasonable we'll try and do a deal. Usually what we call a planning gain deal ...

Planning 'deals' with the retailers are based upon the benefits that a retailer can be seen to bring to a local area: jobs, environmental improvements such as land clearance, or the uptake of a derelict site, or corporate money for associated infrastructure development. This contradicts the traditional view that the retail sector contributes little to local economic development. Such a view has its theoretical foundation in economic base theory, which divides an economy into 'basic' industries which meet an external demand and 'dependent' industries which fulfil the needs of a local market. As an economy is said to need to earn income from outside in order to grow, the 'basic' sector is seen as the engine of growth. 'Non-basic' industries, meanwhile, are viewed as 'residual' activities contributing little, if anything, to local economies. The retail sector has been perceived as falling firmly into this latter category. Thus, conventionally, retailing has been seen by many economic development agencies as a parasitic activity, supposedly reliant upon other wealth-creating sectors of the economy for its wealth and vitality.

However, in recent years this view has changed and the characteristics of its inner-city environment mean that Newton planners have allowed retailers to develop sites that contravene their primary intentions regarding retail structure, because of the benefits that they hope to capture for the local economy, in terms of environmental improvements, on the part of the planners, or job creation on the part of the City Challenge team. Corporate retailers, given the effects of their location and development, have been increasingly seen as core economic activities with regard to employment and town centre vitality. Sainsbury's and Safeway have both been encouraged to develop in this regard, the former in Wellford as part of the City Challenge programme (rather than under the guise of the planning department), and the latter by the planners, to secure the development of a disused and environmentally unsound site. However, on the part of the planners at least, this has involved compromise, and some scepticism remains as to the real benefits of job 'creation':

> Sainsburys is moving up to Wellford. City Challenge were keen to get them there. The conventional wisdom is that these stores employ a lot of people. I say that because, on full-time equivalents, it's nowhere near the numbers they like to quote. Because most people do small numbers of hours. But in the game that City Challenge is in, having said they've created 200 or 250 jobs is a big boost to them. And they won't answer those questions about the impact on the high street shops. And whether or not those jobs are – those types of jobs are very very short hours – it's an old argument actually funnily enough, going around. They are certainly real jobs, I'm not trying to imply that. But um, quality – there is a question that I think isn't asked. The women that are getting these jobs are coming in from, you now, 10 till 12, and then you know, 6 till 8, and I mean, you know, it seems to me that it's mucking up your life. I wouldn't want to work like that. And I don't see any reason why they would. It does allow them to pick up the children, and it does allow them to have three jobs! ... But City Challenge were keen to get them. But in terms of our planning attitude, it was an 'edge of town' centre development, and you can't have it all ways. We say, we don't want you to locate – you know – in the middle of a housing estate. We want you to go to those transport nodal points. That's any of our town centres – we have eight town centres. It's the smallest and has been battered the most over the last twenty years. And in any kind of rational way it doesn't qualify as a town centre anymore. But it is in our plan and so it fits with our policy.
>
> The SavaCentre is going up now ... It's the old gas works. The council entered into negotiations, for a two year period, with British Gas, on a comprehensive regeneration of that site – those two bits were about fifty acres. And it did agree, under, you know, it wasn't our ideal, but to get a major land area like that comprehensively redeveloped. And the argument was that it needed funding from a profitable development, i.e. food retailing. And that's a problem in that it's an 'out of centre' development, but because of the comprehensive nature of that package, securing the British Gas presence located on a partly built site, securing another element of industrial jobs, leisure jobs, and the environmental part, and things like that – the balance of things was that that was acceptable ... We talked to them for over two years to thrash out a comprehensive agreement which allowed compromises on both

sides. And the other one in Mellville just down here ... it was on an industrial site. And we fought that. Both because of the effect on Mellville and the cumulative effect ... and we won that one. So we're not ... the policy is that large stores should locate in or adjacent to town centres – but it's also criteria based, so that there are exceptions – a number of factors – and we will fight some and we will agree some.

From the planners' point of view in Newton council, some of their 'battles' with the retailers have been 'won' and some result in compromise when the perceived 'benefits' of retail development outweigh the fact that the development proposal contravenes plans. Some contests with the corporate retailers, however, are simply 'lost':

The big supermarkets have very slick lobbying operations if they don't get their own way ... they throw a lot of money at it. They always employ planning consultants. If you take the most recent one, the Sainsbury's one; they hired a hall, put on a local exhibition, leafleted all the local people. And said, look, this is wonderful. This wonderful new Sainsbury's. But interestingly enough, local opinion was still a bit split ... we had residents come to the local enquiry, off their own back ... and say we don't like this ... the effect on traffic and the effect on trade were the two main areas of concern ... . In response, when the planning application comes in, we have the council policy to circulate to let people know what it's about. Somewhat wider than we're legally required to do ... so we let people know, but not on the same scale as Sainsbury's. It's a bit biased, because more information is going from them than is coming from us. And in the last five years the change in the nature of the job has speeded up. And at the same time there's just no resources anymore. We were a fairly large department so we could have posed a lot of the interesting questions we've been talking about. I could have argued for a small budget to say, I want to do a before and after study here, I want to ask elderly people where they shop, I want to do a survey of shoppers in Newton to see where they come from and what kind of transport they are using. We can't do that anymore, so a lot of it has to be more speculative.

The dynamism of the food retailing sector, coupled with the growing financial restraints on local government over the last two decades, have made it difficult for the local authority planners to develop a viable response, in planning terms, to corporate strategies. Corporate retailers armed with a battery of consultants have occupied the knowledge gap, creating their own assessments of the local impacts of their proposals to development. This situation is sustained and puts added pressure on the planners:

It's a difficult thing. I mean, again to be quite candid, I – if you've been involved in this field you would have come across the phrase 'retailing is a dynamic industry'. Almost every report I've ever read starts like that. Apart from being a slogan, there is some truth in it. And I do have the feeling that we are lagging behind always ... in planning. In our attempts to respond to these things. So when the major retailers decided to go for superstores and go out of centre ... we were struggling away for ages. Then ... you know, internationally, our discount centres are well below European and American levels, but they're starting to creep in. A lot of us were sort of saying 'well, who are these Lidl' ... and what are the implications of those? The mega-store, the discount warehouse, which is more of an American thing than a European thing. And a third factor that has not really received the kind of response at all is the fore-court petrol station – now you know, just from your own observation, if you pull in for petrol now, all of those are shops now, and collectively must represent hundreds of thousands of units. You could say that that is a return to the neighbourhood store though, you know, they sell a few sandwiches, and coffee and some biscuits and things. So we're not neces-sarily unhappy. But what I'm saying is that's an obvious change and an obvious trend and we haven't developed a response to it because we don't know what it means ...

The discounters are coming in and some are asking for city centre sites and some are not ... we've also currently got seven applications, or rumours of applications, from two of the superstores and five of the smaller ... 10 and 12,000 sort of budget store ... Since PPG13 came out ... we've had one application since they came out for a major super-store. Outside of the town centre, and we turned it down, and we fought an appeal, and we won ... but it's always been our policy to try to prevent them going out of town, even when it was quite contrary to central government ... but it's a confused picture as to whether or not it is more confrontational now than it was before ...

Finally, the inner-city context of high population density and local town centre food retailing need raises particular issues of environmental quality that were not raised by interviewees from the other study site:

The other thing that I'd just like to say that impinges even more directly on your work is that ... the question of road traffic, the articu-lates, the movement of lorries, the pollution ... it's been fairly persuasively argued that quite a lot of that increase of the last ten years is due to the major retailers who control so much of the market, getting both out-of-town stores, and developing ... just-in-time service delivery. So there's deliveries every day instead of the store having a warehouse attached to it. It's got a breaking point, the lorries come in, and deliver to all of them ... It's the 'just in time' method of computer control that's allowed them to do that ... I've heard from Friends of the Earth, who argue – well, they would, wouldn't they – that that applies, but they are suggesting 10–15 per cent of road traffic is down to that factor, and that seems to be very very significant.

Newton borough's food local retail structure appears to have been vulnerable to the full impact of retail restructuring: it was 'hit' by the first wave of food retail change, at a time when planning was at its most open to out-of-town developments. Moreover, given the socioeconomic circumstances of the borough, planners have had to consider the potential for 'planning gain' from retail developments. Despite consistent attempts by the planners to preserve the borough's town and district centres, there has been considerable out-of-town displacement. Alongside the decline in local independent food retailers, some planners agree that this has increasingly marginalised the 'disadvantaged shopper' in an inner city environment of considerable deprivation. These socioeconomic circumstances mean also that the local state must bear in mind other factors when considering retail applications, such as the employment opportunities or environmental benefits that a new development may bring. This necessary priority only refuels the restructuring process, however. In these circumstances, the planners have developed strategies of compromise where necessary, alongside those of opposition and coalition that often have to be employed. Local planners acknowledge their own marginalisation; despite new recent planning guidelines promoting town-centre locations for retail developments, they are unable to fight the dynamic and wealthy corporate sector on equal terms.

Stonefordshire: retail hierarchies for 'middle England'

Within Stonefordshire, the main town centres of Stoneford and Cettleham attract shoppers and other visitors from a large part of the county. Below this level lie the shopping centres of Shington and Tetington. Development in these centres has grown in accordance with increasing catchment areas (associated with more car-borne shoppers) and expenditure levels. Structure planning takes place at the county level; under this jurisdiction, the districts then draw up their own local plans. While planning officers from each district meet regularly to discuss the county-wide approach to such issues as tourism and landscape management, no formal group meets to discuss corporate retailing.

The recent retail geography of the four study sites in Stonefordshire presents a much more mixed picture of corporate retailer–local authority relationships over time. Throughout the county, the tempo of retail restructuring has varied between the urban centres themselves and between rural and urban areas. Corporate retailing is a more recent phenomenon in Stonefordshire. A brief from the Unit for Retail Planning Information, published in 1990, considered the superstore footage/capita ratio for each county in the country, and placed Stonefordshire in the bottom third. The head of Research and Intelligence for the county describes it as a 'catch-up' situation, which has allowed the county to avoid the 'worst excesses' of the 'laissez faire approach of the 1980s'.

Since 1976, only one town centre store closure is directly attributed to 'superstore' development. Since then the county council have 'encouraged' town centre locations, although 'the stores now tend to offer to do that themselves', given the post 1990 focus upon the high street. One key planner described the new policies of the corporate retailers to, conversely, keep their town-centre stores open on an 'eight to eight' basis, because they are 'gold mines', servicing the lunchtime basket shoppers and acting as a 'top-up' store.

Since the late 1980s, when the corporate retailers first began to target the county, Stonefordshire planners have often been able to delay a number of proposals and secure 'priority' locations for new developments, thus confining the siting of new superstores to areas already targeted in the planning structure for development. Since the introduction of PPG6, new development sites for retailers are required to pass a 'sequential' test; the sites that were developed prior to this in Stonefordshire would be able to fulfil these requirements adequately. The largest number of refusals for planning permission on the part of the borough and county planners have been made on environmental grounds, rather than the potential deleterious impact of the proposed development upon the town centre. A senior member of the planning team admits, however, that planners in his county have not had to deal with too many 'silly' applications for planning permission from the corporate retailers; large areas of the county are protected from development because of the official AONB designation of the landscape, and under such circumstances it would be 'ridiculous' to attempt to gain planning permission. In addition, the employment generation argument on the part of retailers is far less forceful than in the inner city context.

The planning departments at the borough and county council level have, however, revised their approach to dealing with corporate retailing planning applications. Originally (in the late 1980s), they adopted what are defined by the planners as 'restrictive' policies, but have moved, like Newton, towards trying to secure 'favourable' locations rather than attempting to fight all developments. For instance, the planners related two occasions when the county fought appeals with Sainsbury: rather than attempting to prevent any development at all, the county went to court to try and secure development on the sites that they had prioritised for development themselves. On both occasions they won. This planning policy of attempting to secure a favourable location for corporate development rather than resisting development completely has, however, often had to be 'watered down' as borough councils have sought to achieve the 'least worst' development site for a superstore. A 'classic case' of this is Cettleham, where four different corporate retailers attempted to get planning permission for different sites.

None of these were favourable to the local planners, but, realising that they would lose if they battled with all four corporates in the courts, they chose to work with the one that posed them the 'least problem'. There are significant variations in these more recent retailing restructuring processes within the county, however.

Cettleham – a laissez faire approach: opening the doors to new retail restructuring

Cettleham is Stonefordshire's second largest town and one of the foremost spas in the country. One of its most important features is the wide range of shops in town, including many nationally known names and a variety of speciality shops. This image has been reinforced by the recent addition of three new shopping complexes. The borough planning department estimates that the town has a 'wealthy' catchment of a quarter of a million people. Those developments that have taken place have not been contested by the planners:

> Cettleham has got a lot of big superstores. There's a Sainsbury's superstore up on the north west of Cettleham, right on the outskirts. There's a Tesco superstore also on the Tetington road, but much closer in to the town centre. And Safeway was built as part of a large housing estate down in south west Cettleham. They're the main three ... There's been no monitoring or surveying at all, but I think it would be fair to say that there have been some closure of small food stores ... In the core shopping area, the focus is very much on *comparison goods*, and that's what a lot of people come to Cettleham to shop for, clothes and foot wear ... I don't know how many food retailers moved out from the core, but er, certainly the predominance of comparison shoppers has been the norm in the town centre for a very long time [emphasis added].

Hence, the level of buoyancy of the town centre is based on much more of a non-food retailing structure of speciality shops and high value durable goods, while food shopping has been located on the edge of town:

> The food retailers tend to remain on the periphery, and they still are there. They haven't all been closed as a consequence of the superstores. But Cettleham has got a large walk-in catchment for its town centre. And quite a high density residential area surrounding the town centre. Which would continue to support food retailers on the periphery of the town centre. I don't know how we compare in that sense, but when we had a recent town centre study conducted by Donaldsons, a firm of expert chartered surveyors and planning consultants, they remarked that that was quite a special feature of Cettleham town centre – this proximity of the residential areas and the amount of walk-in trade that they

found, as the consequence of their survey ... *anyway, we've never refused planning permission* [emphasis added].

Despite its positioning as a centre for 'comparison shopping', discounters have also recently moved into Cettleham. This is intensifying retail competition in the town:

> Recently Aldi have opened, also on the Tetington Road, sort of opposite Sainsbury's. Apparently they like to sort of feed off Sainsbury's, or they like to locate near to another major food retailer. Um, so that actually isn't in our borough, because we've got quite a tight borough boundary. But obviously it works with the Cettleham population. They are the shoppers ... we've got quite strong district centres within the borough. We've got the likes of Solo there, and probably the sort of cheaper end of the supermarket range, things like Gateway, and KwikSave ... There are still local parades of shops all over the borough, on the east side as well. But one of the strong district centres, Coronation Square, is 'bang smack' in the middle of the west side, and that serves a lot of poorer areas.

For the future, the local council is concerned with maintaining the position of Cettleham as the 'premium town centre for shopping for comparison goods', with a focus on fashion and specialist goods. To do this, it intends to begin to take a more active role to promote the town centre, and in so doing, it intends to harness the 'pull' factor of a major food retail development, working *with* corporate power to maintain Cettleham's position in the county's shopping hierarchy:

> The council is a land owner of a large site on the edge of the town centre, called St. James station site. They are just going out to public consultation on some proposals. One of which is to locate a Waitrose superstore/John Lewis store, on the edge of the town centre, on the south west side of the town centre, so the council is actually actively promoting a supermarket superstore at the moment. Waitrose are 'dead keen' to come, but they're getting the public's reaction. Lots of people have said why do we need another supermarket? Why do we need more shops full stop? They point to vacancies in the town centre ... But the level of vacancies in the town centre isn't above national average, and the trend in recent years has been for a decline in vacancies, so that isn't a great issue to suggest that we shouldn't be pursuing this scheme. And a lot of retailers in fact think that the strength of the Waitrose/John Lewis name, would benefit the town centre, because it would reinforce that quality image, and perhaps stop people going to shop in Shington where there's a Waitrose, stop people going down to the new site just off the motorway when the new John Lewis there comes on stream ...

We're at the stage now, having recently had the town centre study completed, we are going to be producing a town centre strategy action plan for five to seven years, in time scale, looking at what we should be doing. Whether we should be allocating sites for new development. Because in the last local plan we didn't do so. Because it was produced at a time when it wasn't foreseen that there was any need to specifically allocate sites for retail development. They seemed to be coming on stream without the council needing to promote development of that sort. Um, but now ... the culture seems to have changed. And with the emphasis on town centres, there seems to be a need for the council to be more proactive in trying to shape the future of the town centre. So I would imagine we would be trying to allocate sites and secure developments.

Stoneford – playing one off against the other

Although Stoneford, the capital of the county, lacks the aspirant finesse of Cettleham, it is also a major centre for comparison shopping. Unlike the latter, however, the Stoneford planners have played a more active role in attempting to shape retail development; and have often had, like Newton, to adopt policies of compromise in order to capture planning 'gains' or 'least worst' scenarios of development. A ring road around the city has encouraged the decentralisation of food retailing, however, and the 'run down' of the city centre food provision has led to an often contested relationship between corporate retailers and the planners. Local authority planners are presently trying to manipulate intercorporate rivalry in order to achieve their planning aims as well as face the increasingly complex embeddedness of the retail land development process:

> we have had a substantial run down on food retail in the city centre proper. And our network of local shopping centres has been denuded ... In 1993 we went through our local plan enquiry. We were faced with a food retail proposal from Tesco who had an interest in a site beyond the Higher Education college. We had a site that we wanted to promote on the cattle market, which offered a number of benefits to us. And to cut a long story short, there are far more detailed involvements, to do with further education, and the potential for a university, and our feeling very disgruntled that the Cettleham and Stoneford College of HE is devolving to Cettleham, selling off our public assets, and so on and so forth – er, it was very convoluted, very bitter, very twisted, and not all to do with food retailing. But basically we had our own 'store war' – the site that Tesco wanted against our cattlemarket site. And a local planning enquiry, which we won ... so they lost that one. So we've now got a site which, at the enquiry, Safeway were involved with the developer on the site, and then pulled out. Tesco were on their high horse by then ... but

we're now waiting for them to develop the site at the cattlemarket ... The problem is that we know that there is another player looking very seriously at Stoneford for another food superstore. Now in the work that was done in the 1993 local plan enquiry, we were able to show through the work that consultants did for us, that there wasn't enough capacity in the market to cope with more than one extra superstore, and that is to be the Tesco store. That's got the benefit of allocation. Anyone else falls foul of that. But we know that a northern retailing interest, not present in the city at the moment, are looking for ten to twelve acres. We do know that they've been looking for someone to do a retail impact assessment ... it is very crucial to us that Tesco build that site, sooner rather than later ... otherwise we're going to have to go back into battle, and we could finish up with a site that we don't want, and that one not being developed at all ...

Stonefordshire planning, like that of Newton, is concerned to protect the food retail facilities available to its most disadvantaged shoppers. Moreover, it is also aware of the problems that the local state has in 'keeping up' with the corporate sector. In this sense, they are nervous about the impact of the discounters isolating the 'deprived, non-car-owning shopper':

We've been getting a lot of interest from discount stores now ... we know that KwikSave as the original discounter is here in the city centre and is looking to spread a network of stores around the city, and the others want to do much the same thing. Now this gives us cause for concern in policy terms. Because we've got this network of local shopping centres that provide for local residential areas ... serving the day to day needs of the area ... the expansion plans of the discount retailers give us cause for concern because we'd need to be shown what the impact would be on these local centres. And if it is detrimental to these local centres, then we'd be looking to refuse them. We've got this hierarchy that's emerged now ... that has superstores spread throughout the city geared towards the car-borne shopper. You've got the city centre stores that are geared towards the workers and those reliant on bus transport to the city centre, and you've got the local shops – basic everyday needs and for those without cars. If the discount stores actually start to operate and wipe out local shopping centres, then the people who would lose out ultimately would be the people who don't have access to cars, or don't have ready access to the public transport system. It's no good to the 90 year old granny on the Bristol road that Lidl and Netto are half a mile down the road, if the shop 200 yards down the road no longer exists. So these are the issues that we now have to look to. And we're trying to develop a policy base that would actually offer that level of protection ... if you've got a car, all is well with the world. So that's where basically we're coming from now. We've got to protect the local shopping centres, and our allocated food retail site. And that's basically where our policy base comes from now … in the context of government advice, etc. …

Tetington: pleased to see Safeway

Tetington is a small rural town that has retained much of its historic and architectural attraction in the three main streets and numerous alleyways of the town centre. Its range of corporate retailing is limited and thus worth listing: Tetington has a large Safeway (on edge of town), a fairly small Tesco, a high-street Somerfield and high-street independents. The little development that has taken place has been uncontroversial, although the impact of the recent development of Safeway was of concern:

> We haven't had any controversy with the development of food retail sites … the Safeway went up on the edge of town before the housing that is planned for that site was built, but that's started now … as far as the impact of these developments go on the high street, the Tetington one is instructive … because it's the only one we've got comprehensive records for. We've done shopping surveys in the town centre both before and after that's happened … in September 1993 it opened … and what's actually happened, a number of empty units in Tetington town centre has actually gone down since the Safeway opened. But … it's not quite as straightforward as you might think, because what's happened, is we've introduced a small business grant scheme, to try and encourage people to take up units, so they can get grant aid for opening units. So that muddies the figures … the independents were struggling anyway before Safeway arrived. They were particularly concerned about that. But in the end it was OK. There wasn't really strong opposition to that Safeway … it wasn't too bad … because it's fairly near to the town centre – about a seven minute walk to the town centre … And there was the possibility that it might even attract people into Tetington to do their weekly shop, who might otherwise have gone to Cettleham … A couple of butchers have gone in recent years, and a baker, but the greengrocers still seem to do ok.

Outside of the town of Tetington, however, the pattern of food shopping in the borough's rural villages has changed quite dramatically over the last fifteen years. The county as a whole has actually experienced an *increase* in the number of outlets selling food, but this is because of the emergence of farm shops and 'speciality' food shops selling 'upmarket' produce to the car-borne customer. Alongside this has been a drastic decline in the number of village shops selling groceries. As some settlements are served only by weekly bus services, access to food shops for poor rural people has now become an issue of 'serious concern' for the county council.

Shington: catching up

Shington, as with each of the other study sites, has experienced the local articulation of the national restructuring of food retailing in a unique way. The Shington experience is described by local planners as a 'catch-up' situation, since major developments have only occurred since the late 1980s. While there has been a 'decline' in the town centre since then, in terms of the loss of independent food retailers, this has been complicated by its concurrent expansion and up-marketing. Since the first corporate retailer arrived in 1989 there have been a series of 'store wars', as the local authority have revised their strategies towards development. Echoing the concerns of the Newton authority, Shington seems to have realised only in hindsight the implications of the new configuration of retailing, and is now actively attempting to ameliorate it. Moreover, Shington has, like the borough of Tetington, experienced a decline in traditional rural grocery provision. Because this story is recent and compact, it has been well-documented and is therefore told in detail here:

the decline of the town centre has certainly been a factor in recent years, and I suppose, in the recession years of the 90s, that's been a trend that's recurred and unfortunately stuck with us. I think it's been exacerbated to a certain extent by the fact that the new developments schemes that have taken place in the town centre – there were really two of those. One of them is the Cornhill development, which is to the east of the town centre – and it's got twenty small units there. They're not food based, they're just general comparison goods stores. That one was completed in about 1990. But of course that came at the wrong time, it came at the end of when things were good for the retail sector. So since that time, that particular part has never been fully let. So that part always looks a bit vacant. Also on that side of town there's a much smaller scheme which is a council scheme. Which was put together roughly at the same time – eight units – but again, that one's never been fully occupied ... So it's not just the decline of the high street but an expansion that didn't get filled. If you go back say ten years, in the mid-80s, then what you have, in terms of food anyway, is a very typical pattern. You have three smallish food stores in the town centre: Gateway, International and Fine Fayre. There were no big supermarket chains at that time anyway. And we did a survey in about '85, quite a big retail survey, just asking local residents and shoppers to the area what they felt about the town's shopping facilities ... and what they were looking for was ... were for a named store to come in. Local people were ... there was a very strong emphasis on it. Because at that time, Tesco had just opened up south of Stoneford. So I think that was in people's minds. So typically people were saying that they wanted one or other of those types of stores. That's what actually happened in the intervening years, but they haven't actually come right into the centre, although Waitrose is just on the edge of the town centre.

So Tesco were the first to come. Just outside the town centre, in about 1989. Certainly, we tried to be as helpful as possible because we felt there was a need to create more shopping floor space, that's from the planning side of it anyway. And that was then backed up by the type of surveys we'd done a few years earlier ... there's a whole story connected with that scheme ... trouble with the trees, and all the rest of it ... I think it was really looking at the wider area, so we were looking really at the Valleys, because something like that doesn't particularly help the town centre I don't think, because what then happened is that the other two stores, International and Fine Fayre, closed soon afterwards ... The only one left now is Gateway ... When Waitrose came a few years later in 1993, they're more or less in the town centre, so that site was always one that was more beneficial to the town centre ... but it took a long time to be taken up because it was a fairly tight site and needed quite a lot of work doing on the design ... But certainly at the time, I don't think that we had any problem with those two developments.

It would probably be fair to say that the high street grocers have taken a bit of a back shift in recent years. But there has been a sort of upgrading of the town centre if you like, since about the mid-80s, if you went to the town centre ten or eleven years ago, and walked down the high street, you'd find a number of the units in a very poor state of repair. It was only a couple of years later, that things started to move. There were no redevelopment schemes in all those years ... by doing that, better units and things, better companies came in: people like Boots, so high names did start to come in where they hadn't before. And we've got more upmarket clothes shops. Although I take your point that people like greengrocers, butchers, bakers, those sort ... they haven't benefited ... but we haven't been able to do another shopping survey ... Also, what's happened is that you've got centres like Stoneford, fairly close by, expanding fairly rapidly in the provision of their town centres. Particularly in food stores. I think that's been quite important too ...

There's been a lot of controversy because I think we've been in a sort of catch up situation. Shington's taken a while to catch up. It's a second tier, I would have thought, to the city areas, because the population's much more diverse. Shington itself has got quite a large shopping catchment area. The population is about 104 or 105,000, I think. So you've got potentially quite a big demand.

At the time of the interview, the planners had recently lost an appeal to Sainsbury's:

Now a Sainsbury's is going to be built out at Dudbridge ... just across the fields. It was an appeal decision, so it's not something we supported, because of the effect on the town centre ... What happened, in a strategic sense, there was a requirement, or a need identified, for another store. Developers obviously got hold of this, and started looking at this place in about the early '90s for another site ... not to do with the size of the population, but commercial need. Well, demand created by the local population, but yes, it was more commercial ... in terms of the profitability of operating a store ... But what we didn't want, particularly, was the site where they were looking. We wanted one much nearer to the town centre. Because the one that has been allowed is ... maybe even three-quarters of a mile from the town centre ... but the Department of the Environment agreed it. They all (the different supermarket chains) came in, and turned the lot down. There were about six different sites. Several of the stores were involved in the sites around – um, and in the end the Inspector plumped for that one being, I suppose, the lesser of the evils ... We had refused all the applications. There were then a couple of very large enquiries – comparing the different sites, and that one, to the inspector's mind, seemed to be the most appropriate.

The key to the planning permission for the site for Sainsburys was the Ebbley bypass – because they felt that these were areas that could be very well serviced ... to a certain degree it was taken out of our hands. Obviously, you argue the case that you put forward, when the planning application makes committee ... and in a sense the Enquiry is actually just going over ... just building up the case from those points ... I think we were resigned I think to losing it ... because the strategic picture backed up the need for another site. So we didn't have that argument – it was something I think that was accepted. That there was a need in trading terms for one more site ... Certainly the supermarkets, around the time the appeals were going through, were producing very glossy brochures, which were aimed really at our members, members of the planning committee. Presentations and things like that ... it's very much a selling job ... particularly to councillors on the planning committee who were going to make the decision ... they're the ones who actually make that decision ... and so you find a certain amount of lobbying goes on ... they'll try and persuade people and things like that ... but the councillors supported the planning department and the town centre and tried to prevent further out-of-town development.

A lot of work is being done currently, in connection with a community planning conference which is being set up, which is specifically aimed at trying to – in Shington – take action at the grass-roots level – about the problem of the town centre. It got going last year. The plans haven't come out yet, but the idea is to involve the community in its widest sense. Involvement of traders, people shopping etc. To come up with ideas about what can be done to help the vacancies ... but because we're a district, we've got a number of other centres, like Stonehouse, Aylesworth, Dursley ... and at the same time we've got to try and help those centres as well. But the council, in recent months, have given priority to trying to reinvigorate the town centres, and have identified quite a large amount of money for a regeneration fund, improving the physical characteristics, the shop fronts etc.

We've had some interest from the discounters, but no planning consents granted. Nobody has really even taken it to application. As far as the big supermarkets go, there's still interest. For instance, there's interest in the south of the district, in Dursley. We've got a site there, which is tied in really with the whole regeneration of the town centre and ... we've done a study fairly recently on what needs to be done there ... and part of that is a site for a new food store ... it's really key to a lot of other improvements that are needed in the town centre ... There's been talks with several of them ... we've got Tesco, we've got Sainsbury's as well ... to be honest, they often come to you because the word gets out ...

In this case of a smaller town, the more recent 'cascade' of corporate retailers down the urban settlement hierarchy (with the large corporates capable of generating profits from smaller stores in smaller locations) means that the process of retail restructuring in such local spaces is far more recent. Moreover, the incorporation of the dynamic corporate retailing sector into broader and progressive regeneration and redevelopment schemes for such firms gives added impetus to the retailer-led domination of (and insinuation in) urban settlement patterns, as well as the further removal of smaller independents. Redevelopment schemes, once completed, are however somewhat at odds with their associated programme ideology and character, and are increasingly typified by higher small shop vacancy ratios. The planners' dilemma becomes one of how to solve these vacancy problems, having given their blessing to the reconstitution of town centres by the corporate retailers.

Conclusion: the organisation of food regulation

Changing retail space

The London borough of Newton has been a longstanding dynamic site for food retailing, and has experienced local restructuring from the 1970s onwards, at the same time as corporate retailers themselves were restructuring. This process has been concentrated and explicit in the context of the inner city. It illustrates the 'classic' impact: the loss of the relative importance of some of the borough's shopping focal points, as edge-of-town and out-of-town superstores have been built, and a large loss of independents throughout the borough. However, Newton planners have managed to maintain Newton town centre. Planning policy has been based upon the ideas of equitable provision and access to food for the borough's population; however, this is tempered with a need, because of the socioeconomic circumstances of the borough, to compromise for planning 'gain'. Retailing has also been viewed, within the ideological context of 'City Challenge', as a valuable source of employment. Post-1990 planning policy developments (such as PPG6) may have ostensibly given more power to the local authority; however, planners see themselves, in the face of corporate power, as capable of little more than 'bucking the trends'.

Stonefordshire, conversely, with its different regional shopping centres, represents a variety of different local situations and retailer–planner compromises, as the pace and extent of change has varied across the county. Stoneford has experienced change on a long-term basis that has recently involved contestation; Shington has experienced change only very recently, but that has also been contested. By contrast, local authority planners in Cettleham and Tetington have had little conflict with corporate retailers, instead viewing the developments that have taken place as being of overall gain to the local economy and town structure. Stoneford and Shington have both developed strategies of compromise with the retailers, as they have adopted proactive strategies in an attempt to ameliorate the consequences of retail change upon their marginalised shoppers and retailers, respectively.

These different depictions of retail restructuring and local hierarchies of retail contestation developed in the two study areas are clearly the result of the interactions of local state policies and the spatial manifestations of the competitive restructuring of the corporate retail sector. These dynamics produce spatial outcomes which structure the study areas as hierarchical and tiered food consumption spaces. They provide the emerging geography of consumption space on which local food regulations have to adopt their strategies of regulation.

We can now turn to the local organisation of food regulation in the study areas. While the planners, officials and committees may, as we see, shape to varying degrees these dynamic retailing development processes, it is another set of public professionals (EHOs, TSOs and Public Analysts) who are bestowed, by the state, with the responsibility to regulate the actual foods which flow through these retail outlets. It is they who have to cope with the retail hierarchies developed in these food consumption spaces.

Professional space

The responsibility for public food regulation is assigned by law to two professional groups: Environmental Health Officers (EHOs) and Trading Standards Officers (TSOs). Corporate retailers employ both types of professionals for in-house regulation, but commercial food premises of any kind are inspected and monitored by officials employed by the local authority. Within this setting, however, both EHOs and TSOs have a remit much wider than food regulation. Local authority environmental health departments are responsible for maintaining local environmental quality in its broadest sense, within and outside the home and in commercial premises. Officers enforce legislation concerning such things as health and safety in the work place, noise and local environmental pollution control, and the maintenance of public housing. Trading Standards Departments, conversely, are concerned with monitoring the legality of any commercial enterprise within the local authority for the protection of the consumer and honest trader, investigating such things as the legitimacy of advertising, consumer credit or the safety of second-hand cars. Public analysts also have a role in food regulation but they too have a much wider remit, being involved with the chemical analysis of non-food substances and undertaking safety checks on such things as children's toys.

Fully qualified EHOs have attained accreditation from the Chartered Institute of Environmental Health, either by following a degree course in environmental health or by undertaking postgraduate study (older officers may have qualified without completing a degree). This training covers all aspects of the profession's responsibilities. However, officers do generally specialise within local authority departments. Although the internal structure of each local authority was not uniform, all of the five that were involved in the study had small specialist food teams of between four and eight staff, within an overall environmental health function of perhaps thirty staff (specialist 'housing' and 'pollution' teams were also common).

The Trading Standards Departments, however, did not show the same degree of specialisation. In the unitary authority of Newton, the trading standards team was split into two on a geographical basis: one team was responsible for undertaking all trading standards activity in the west of the local authority, and the other was responsible for the east. The latter team, involved in the study, comprised eight members of staff. It was estimated that 50–60 per cent of their time was spent on food issues. The trading standards function in Stonefordshire was also county-wide (trading standards always operates at the county level, regardless of unitary or non-unitary status; in consequence, there are many more EHOs than TSOs). This however, was a little more specialised. A specialist section of six staff, from an overall number of about thirty-five, were responsible for food retail premises only. Fully qualified trading standards officers have attained a certificate from the Weights and Measures Authority, giving them the right to enforce legislation in this area. In recent years, this has generally been gained as a postgraduate qualification, after an individual has been 'taken on' by a local authority for in-house as well as external training. The law requires that each local authority appoints a Chief Inspector of Weights and Measures, who is generally also the head of its trading standards department.

Within the Environmental Health and Trading Standards Departments, there is a distinction made between fully qualified officers and technical staff. Generally, the fully qualified officers comprise 50 per cent or less of the team. All the environmental health food teams included in the study areas have two or three fully qualified officers, and they have a similar number or slightly more technical officers in their staff. Qualified officers are entitled to individually enforce all aspects of the food law: to serve 'notices' when necessary, which legally require commercial premises to improve their standards immediately or be prosecuted; or, on a rare occasion, to serve a prohibition order to immediately stop a business trading if it poses an imminent serious health risk to the public. While most technical officers are trained to carry out all the duties of a fully qualified EHO, they are not entitled to serve notices or prohibition orders, and would have to seek the assistance of one of their qualified colleagues in a situation where legal action was required. More recently, provisions made under the 1995 Food Hygiene Directive have also disallowed technical officers to inspect 'high-risk' premises unless they have fulfilled the requirements set by the Institute.

Staff within Trading Standards Departments also have varying qualifications and thus professional entitlements. In Newton, only two members of the team that were studied were qualified to enforce Weights and Measures legislation, although the technical officers were trained in-house to carry out inspection duties. In Stonefordshire, in addition to the two fully qualified officers and other technical staff, the food retail team also included two school-leavers who were employed as general office assistants, with a view to upgrading their responsibilities through further training.

Food regulation is undertaken by both Environmental Health and Trading Standards Departments, but central to the operation of law enforcement in this area is the boundary between the two professions. Their respective concerns with food reflect their wider responsibilities. Very broadly, environmental health is concerned with the protection of the consumer from foods that are hazardous to health, through the enforcement of hygiene standards. Trading standards, conversely, focus upon the protection of the consumer from unfair trading practices such as short-weighting and deliberately misleading food labelling in relation to food composition. With regard to commercial food premises, therefore, they enforce different aspects of the law.

Both, however, undertake their responsibilities in a similar fashion, having several distinct sources of work. First, routine site inspections are the most important aspect of their work and the most time-consuming. All the departments have implemented, over the last few years, a computer-generated hazard rating for each food outlet under their jurisdiction, from which the frequency of the need for inspection is assessed. Inspection 'targets' are set by MAFF for each Environmental Health Department. Second, both EHOs and TSOs have to respond to complaints received from the public, as well as issuing them with advice in response to enquiries.

This work has increased in recent years due to increased public awareness of food safety issues. 'Sampling' activities also constitute a significant part of the work for both professions, where certain food products are targeted in either a nationwide, regional or local campaign for random testing. Educating local businesses about new legislative developments is time-consuming for both EHOs and TSOs; many Environmental Health Departments also run food hygiene training programmes for the employees of local food businesses. Finally, as we shall discuss in more detail in subsequent chapters, the Home Authority Principle generates further work for some local authorities. This principle dictates that any food company that trades outside of the local authority where its head office is located can name one particular authority (normally that one where its head office is located) as its home authority. The home authority is then responsible for liaising with the company as a whole, and dealing with any complaints passed on to it from other authorities throughout the country.

EHOs working within specialist food teams also have to undertake non-food duties. Each inspection of a food premise also includes a health and safety inspection (apart from very large premises which fall under the jurisdiction of the Health and Safety Executive). Departments also check planning applications for prospective commercial food premises to ensure that they will comply with legislation. Other responsibilities are variably added: most are responsible for the monitoring of infectious diseases (not just those spread by food such as salmonella and E. coli); and some, such as is the case in Shington, are also responsible for overseeing public burials. Shington is also the only food authority in the country that continues to undertake meat inspections, since the Meat Hygiene Service took over this role in 1995 from all other Environmental Health Departments.

Environmental Health Departments all liaise with both MAFF and the Department of Health regarding new legislation, and have to report their inspection and sampling activities to MAFF on a quarterly basis. Trading Standards Officers liaise with the Department of Trade and Industry and the Weights and Measures Authority. LACOTS has a coordinating role between all local food authorities, working to advise local authorities on the interpretation of the new legislation and to attempt to ensure its even implementation across the country. Local food authorities also communicate with each other electronically. In addition to this, the non-unitary authority of Stonefordshire has a quarterly county-wide meeting of its 'food group', where representatives from each of the Environmental Health Departments, as well as the Trading Standards Department, get together to discuss matters concerning food regulation. In the unitary authority of Newton, however, there was little liaison between the environmental health and trading standards functions. Regional groupings, such as the West of England food group, also encourage wider liaison.

9 The nationalisation of food regulation

Coping with space

Introduction

Earlier chapters have outlined the substantial developments that have taken place in British food retailing over the past two decades, and have examined their repercussions at both the local and national level. First, we have seen how the pattern of food provision has evolved in cities and towns and rural areas, as corporate growth and restructuring has been manifest both in changes in the size and physical distribution of corporate food outlets and in the decline of the non-corporate sector. While these changes have been nationwide, they are witnessed and experienced locally, and as Chapter 8 described, are shaped by the particularities of locality. Second, at the national level (or more specifically, at the level of national government and corporate national head office), we have explored the revised relationship between the state and the superleague retailers (see Chapter 7). As we have described in our model of food regulation in that chapter, the ascension of corporate power has also coincided with a process of state restructuring, and has led to the emergence of a 'private-interest' style of regulation that contrasts with the traditional regulatory style based upon notions of the public interest.

For several reasons, however, as we have seen, these retailer–state relations have not been straightforward or by any means static. Indeed, they have to be reconstructed constantly. For example, concerns over food safety and the Conservative government's rhetoric of deregulation have raised contradictory impulses and thus (potential) tensions within the regulatory framework. At the same time, the consolidation of the Single European Market has brought forth an unprecedented flow of regulations relating to the food sector which the government is legally obliged to implement. Implicit in this, and central to it, have been tensions between (national) policy and (local) implementation. It is to this that we now turn our attention in this chapter, and in so doing, we map out the intricacies of the broader political–economic shifts that have thus far been documented. This shows that despite the variations in retail restructuring and food provision across difference spaces, there has been a strong attempt from both national government and the retailers to nationalise the food regulatory system. These tensions – between national systems of regulation and the vagaries of local implementation – have been a particular characteristic of food regulators of the 1990s. Moreover, they have been exacerbated by the contradictions associated with the government's growing public responsibilities and pressures in delivering food choices at a time when its powers have been devolved.

One of the key dimensions of regulatory uncertainty has surrounded EHOs and TSOs. The professional role of EHOs and TSOs has been contested and then adjusted to accommodate and 'catch up' with the revised regulatory relations between the national state and corporate retailers. During this process, local-level food law officials can be seen to have 'lived out' the internal tensions of the national relationship. As the tensions continued, it has been those responsible for administering and implementing policy who have also had to carry much of the weight for its re-evaluation. In the following analysis, we depict the renegotiation of the professional role of EHOs and TSOs as a dialectical process between actors placed along the local–national continuum. We show how this has led to the introduction of new codes of practice and procedures for the enforcement strategies of these groups, and, in theory at least, a homogenisation of regulatory practice across the country (or between local food authorities). Thus, through a series of stages, the outcome of the revised 'private interest' retailer–state relationship has been to promote the nationalisation and aspatialisation of local level food regulation, across the 589 local food authorities in the UK.

Below, we explore in detail the different features associated with the nationalisation of food regulation suggested by our model outlined in Chapter 7. Here we describe the ebb and flow of implementation strategies. The period immediately before and after the turn of the 1990s proved to be formative and provided the backcloth against which our fieldwork in 1995 and 1996 took place. When the relatively benign regulatory environment to which the large food retailers had become accustomed was fractured in the late 1980s by a series of food scares, the response within government was a heightened sense of the need for public-interest regulation. The result was a perception amongst many of the food retailers that the rules of the regulatory game had been changed, and unfairly so. A more onerous regulatory style for the retailers inevitably provoked a reaction. This in turn led government to modify its advice to EHOs, such that once again they adopted a more conciliatory role towards those they regulated. We present further evidence from our two case study areas in examining these regulatory changes.

Contested regulatory domains: local/national interactions (1990–5)

The emergence of tiered regulation

Between the passing of the 1990 Food Safety Act and the UK implementation of the Food Hygiene Directive (Council Directive 93/43/EEC) in September 1995, the contestation that ensued in food regulation needs to be explored in its component parts. First, there was the *regulatory advance*, involving new powers for local enforcement officials, such as Improvement Notices, which, if issued by an EHO, legally obliged substandard food premises to undertake the action outlined or else face prosecution; and the regulatory ramifications of the consolidation of the Single European Market in 1992. Second, there was the *regulatory retreat*, as officials' more stringent actions attracted the concern of the major retailers and also of parts of the media, encouraging the government to embark on a 'deregulatory' drive. Third was the *rapprochement*, aimed to assuage the interests of government and major food retailers; from 1993 onwards, new codes of practice and other regulatory tools were introduced. As public-interest and private-interest practices were revised to solidify a tiered system of differentially regulated food retailing, new enforcement practices incorporating ideas of hazard analysis soothed the tensions along the local level public-interest–private-interest regulatory interface.

The regulatory advance

Hutter (1988) has noted the importance of public and political interest in influencing the enforcement strategies of Environmental Health Departments. Undoubtedly the food 'scares' of the late 1980s, such as those relating to listeria and salmonella (see, for example, Smith 1991; Lacey 1991), along with growing concerns about BSE, heightened both political and public awareness nationally and locally to food quality issues. As the saliency of food quality increased, so EHOs shifted their regulatory strategy from an *accommodative* stance – that is, one which is 'co-operative and conciliatory' (Hutter 1988: 6) – to one that was more *insistent* – in which there was a quicker recourse to legal means to secure compliance (Hutter 1988: 156). As local food law enforcement officials, they had to assuage both political and public disquiet on an issue that had previously attracted little media attention.

EHOs, adopting a more insistent approach, brought a new zealousness in the public interest into their everyday discretionary decisions. In doing so, EHOs were assisted by new powers under the 1990 Food Safety Act and an Audit Commission (1990) report that demanded more intervention by previously quiescent environmental health departments. As a senior member of LACOTS reflected:

Now, in the late 1980s, we had all the food scares and everything else which really took food to the top of the agenda. The government said – it's going to provide new legislation – the Food Safety Act – and deal with these issues ... of course, what the government was saying was that we need stronger enforcement powers ... The Audit Commission was saying that their survey confirmed that there were appalling conditions out there, and that enforcement was not working. And that Birmingham City Council do a wonderful job – they get out there and enforce the law ... they close places down, they prosecute people and it's effective ... So we got new legislative powers [under the Food Safety Act]. The two key things were the [Improvement] Notices and the streamlined procedure for closing premises down. We were provided with statutory powers that really worked ... in the past if you did an inspection, and you found something wrong, you essentially had two options: you either dealt with it informally – you wrote them a letter – or you prosecuted ... Now an Improvement Notice is basically a statutory notice which says 'you are breaking the law, and you have to do this and if you don't do it, you're committing an offence'. So it was like a sort of a step up of forcing people to do the work. Government, well the Audit Commission, were very positive about enforcement ... I've kept my notes and my notes say get out and use these powers. So EHOs did that, they felt that that's what the government was telling us to do.

EHOs were informed of the new legal requirements via formal letters, circulars and bulletins from the Chief EHO at the Department of Health, and new Codes of Practice issued jointly by Department of Health and MAFF. The message, as the Chief EHO from Cettleham explained, was consistent through the hierarchy:

I have Codes of Practice that were issued under the Food Safety Act, I have Audit Commission reports, and Audit Commission performance indicators where it was very clear that the government came out and said to local authorities we have provided you with some very tough powers indeed, go out and use them. Okay. A lot of authorities went out and used them. And there were very clear directions from Mike Jacobs then, who was the Chief Environmental Health Officer at the Department of Health, very explicit that came through in the first Code of Practice number five, that improvement notices shall be considered as the first option on each and every occasion ...

The Codes of Practice issued from 1990 onwards under the Food Safety Act (each covering a different aspect of food regulation, such as the use of Improvement Notices or the enforcement of temperature legislation) deepened the links between national and local actors in food regulation. They formed an essential tool in the revision of food regulation. As the Chief EHO from Newton explained:

> The Codes of Practice that are produced by MAFF are vitally important for us to read. We have to read them. They're written in bold or in ordinary script and bold script we've got to take notice of. They say that local authorities are expected to carry out the recommendations written in bold script, and if they don't then they've got to show good reason why they aren't. The other light script ones are recommended to be taken up by the local authority but there is some choice – they are sort of a bit more voluntary. We use these things for guidance ...

In short, the powers that EHOs were now to wield were quite clearly to be used to ensure that baseline food standards were maintained and thus that the public could be safeguarded in the food that it consumed.

While the passage and content of the 1990 Food Safety Act had been influenced by the food scares, and its content in part shaped by the nature of subsequent debate, its formation had also been affected by the need to comply with European legislation. Indeed, the period of the creation of the Single European Market saw an increase in directives and regulations from the European Community, a number of which were concerned with the harmonisation of food law between member states. These in turn had to be implemented in the UK. According to Booker and North (1994), outspoken critics of food legislation, this created a 'regulatory explosion' in 1992–3. Following in the wake of the 1990 Food Safety Act, they served to further increase the degree of public-interest regulatory activity undertaken by environmental health officers.

Finally, during this period of regulatory advance, with government believing that there was a need for increased direction and coordination amongst EHOs, the remit of LACOTS was expanded to accommodate food hygiene as well as trading standards concerns. From 1991 onwards, LACOTS took on an advisory role for local authority environmental health officers, providing written and verbal guidance on enforcement that was in addition to that issued by the Ministries through the formal Codes of Practice. Moreover, as a senior member of staff explained in interview, they immediately set out to improve 'effective networking' across the country in order to 'establish a national perspective' on enforcement practice.

Protest from the superleague and small food businesses: the regulatory retreat

By 1992, both the top tier of food retailers and the smaller independent food businesses were registering complaints about what they perceived to be the unnecessarily harsh, damaging and uneven local-level implementation of food law by enforcement officials. Meanwhile, within the media and amongst

Members of Parliament could be heard cries that European legislation, together with the new powers accorded at the national level to EHOs, were having a harmful effect on British small food businesses.

The large retailers, enjoying close links to government, were at least as well able to make their particular views known. As a senior official from the Department of Health explained:

> at that time we were getting a lot of complaints from industry about EHOs. It was all rather unfortunate really, and the Ministers weren't very happy. It was a combination of things. First of all the Act, and then the whole lot of EC legislation coming in, resulting from the completion of the Single Market. It got the industry a bit paranoid thinking that EHOs were going to come round and do lots of terrible things. And another factor that was a bit unfortunate was the Audit Commission report which did rather over-emphasise the enforcement side of things. It rather encouraged them to be rather more hard headed over enforcement – giving them the impression that they should be rushing around giving demonstrable evidence of enforcement activity ... and of course industry didn't like it and they made a lot of noise ...

The 'noise' was listened to. By the end of 1992, new devices were being devised at the national level to revise the implementation of public-interest regulation and its interface with privately regulated interests. This then resulted, in policy terms, in a system of food regulation that clearly differentiates between the 'superleague' and the rest of the UK's food retailers. The 'superleague', with retail outlets located throughout the country, were able to identify variability in enforcement practices between food authorities. From their perspective, it was imperative that national government devised legislative tools to nationalise regulatory activity alongside retail change.

Towards a rapprochement

From 1992 onwards, in order to placate the disquiet from the retail trade, new policy initiatives designed to modify the food law implementation practices of EHOs were introduced by government. The first of these was to return local-level enforcement to a more persuasive and traditional style of operation. A LACOTS interviewee explained the shift in emphasis as follows:

> The government had been encouraging enforcement, and then the [Audit] Commission obviously were doing these detailed directives and people then started having question marks ... And so EHOs are criticised ... And then suddenly, the government then decides deregulation is the day ... and suddenly says we don't want you to serve [Improvement] Notices now. And there is European law, but we don't necessarily want you to enforce the European law ... the deregulation

thing has impacted quite heavily on how we attempt to implement European law.

During 1993 and 1994, LACOTS issued a series of Guidance Notes to EHOs working for local authorities across the country (LACOTS 1993; 1994b). These outlined the new principles for the implementation of food law, encouraging them to use their discretionary enforcement powers in a 'gentler' fashion. For those local level enforcers 'living out' these tensions at the national level, this required another adaptation to their professional role. As the Chief EHO from Newton explained:

> After 1990 we served a lot more Notices, and proved that when we served a Notice we did actually mean business. That's what MAFF had wanted us to do ... But I think MAFF got concerned about some of the particular aspects of this ... and they advanced new advice to us, that when we serve Notices we only serve them if the people are not complying or have a history of not complying with informal letters. Which means that we have to start building up a profile, of the premises that do comply and those that don't ... it means that the enforcement is a little bit more woolly, like it was in the old days. And what's also clear is that we're not serving anything like the number of Notices. In fact, they've plummeted from hundreds down to single figures.

Similarly, an EHO in Shington described the situation as a pendulum:

> the pendulum had sort of swung all the way to the point where the government can say – you know, there's a lot of food businesses out there, trading in unhygienic conditions ... local authorities are basically being ineffective about it. This is as a result of an Audit Commission report back in 1989, or something like that ... and really you've got to get your finger out and get on and sort them out. I think that's what a lot of local authorities did. They had new enforcement powers, they had the power to serve improvement notices or prohibition notices, and we used them. And unfortunately, it was at a time when recession was biting, and businesses were feeling the pinch. And I don't think they liked the reaction that they were getting from the local authority, probably at a time when they felt they needed understanding and time, to sort themselves out, the local authorities were saying look, you know, the legislation's quite clear, this is what you've got to do. And if you don't do it then we will serve improvement notices ... And now of course it's gone full swing, and we're back to the situation where we play down the enforcement function, and educate people more.

Finally, the following extract from the field diary explains the impact of this changed government emphasis upon the professional role of an EHO in Tetington:

> J ... only joined the EHO team here in the last few months ... She said that even in the few years (post 1990) since she trained, she is discovering that the role of the food law enforcer has changed. She thinks this is largely to do with de-regulation. She says that now, she sees her role very much as that of an educationalist rather than an enforcer ... In the big stores now, for instance, they are not even expected necessarily to actually undertake an inspection of the premises. Instead, they are just meant to have an interview with the management about the systems that they have in place to control their hazards in their food systems. Before this, a visit was about checking the walls, the floors, the ceilings etc. Now it doesn't matter so much if the surfaces aren't perfect, as long as they have identified their hazards.
>
> She went on to say that the regulations have changed the role of the enforcer very much, and put more responsibility onto the shop manager. Apparently though, at the individual retail outlet, this is often an unpopular role. For example, it used to be the rule that every food premise must have a nail brush. Now it's not, but the shop must identify its own requirements to control its own hazards. Yet many shop managers ask her to just tell them exactly what they need to do to comply with the law ... We commented that the directors of large businesses may lobby government to change the rules so that they are left to their own devices; however, at the local level this is not always popular, especially when a retailer knows that they must be able to demonstrate due diligence.

In addition to this change of emphasis, the above extract hints too at the new tools that were also introduced by national government to change the way that local level enforcers operated. These had the effect of bringing EHOs and TSOs 'in line' with national developments, allowing them to practice differential enforcement between the privately regulated 'super-league' and the second tier of food retailers. Moreover, in theory at least, their implementation should result in a spatial uniformity of enforcement across the country in relation to the nationally organised corporate retail geography. It is to a consideration of these new regulatory tools that we now turn.

The nationalisation and aspatialisation of food regulation

The introduction of new enforcement tools – the home authority principle, industry codes of practice and techniques of hazard analysis – have served to unify enforcement practices between local food authorities. Below we outline these, and pay particular attention to hazard analysis, which has gone furthest in allowing local regulation to traverse the tiers of food retail and differentiate between publicly and privately regulated retailers, in line with the revised relationships at the national level between government and corporate retailing.

The home authority principle

As we have already seen, in the early 1990s retail companies had alleged to government that various store managers throughout the country were relating very differing accounts to Head Office regarding their treatment at the local level by food law officials. Partly to counteract these complaints, the co-ordinating role of LACOTS was expanded to incorporate EHOs, who were then themselves obliged to incorporate the *home authority principle* into their work (LACOTS 1994a). Under this provision, nationwide food retailers are entitled to elect one local authority (usually the one in which the company's head office is located) through whom liaison with local government regulators will take place. While each individual store can be inspected by enforcement officials from its locality, any complaints must be channelled via the inspecting officer to the Environmental Health Department in the home authority. It is presumed, therefore, that a working relationship will be established between the relevant authority and the retailer, thus deflating earlier tensions along the interface between public and private regulation. The home authority principle is endorsed by the major retailers. As the British Retail Consortium has commented, 'Local enforcement is appropriate, but the Home Authority principle should be encouraged and national supervision and co-ordination is necessary to guarantee a given standard' (quoted in James 1997: 43). Similarly, Tesco believe it would be 'advantageous to have central co-ordination with respect to national multiples, an enhanced home authority approach' (quoted in James 1997: 45).

Industry codes of practice

New procedures and techniques of assessment have been introduced into the work of EHOs and TSOs by the Food Hygiene Directive (Council Directive 93/43/EEC), which was ratified in September 1995 and incorporated into UK food law. The first of these are industry codes of practice, allowing the food industry to develop its own guides of compliance with food regulation. Different sectors of the industry have produced their own guides; for instance, there is a guide for the catering industry, and another for bakers. These appear to have been met with a favourable response from local-level enforcement officers, as the Newton EHO explained:

> It's rather along the lines almost of self-enforcement. If they [the food industry] produce their own guidelines then you can enforce the law according to their guidelines that they have written for themselves. Otherwise we have to make up the guidelines for them. The industry goes away and thinks up some decent guidelines which they think are going to be fair for their members and reflect particular problems that are part of the food industry.

What is significant about these codes of practice for our own model of private-interest regulation is that companies are forced to pay greater attention to their supply chain relationships. The issue is no longer simply regulation of individual premises, but the management of risks along the supply chain. The management of risk or hazard has now become a cornerstone of food regulation.

Hazard analysis

The hazard approach has two components. First, the frequency of site inspections by EHOs is decided by the food safety 'risk' posed by the business, which is assessed according to a four-factor equation based upon the perceived turnover of trade, the type of production involved (dependent upon the nature of the food product made or sold and the risk associated with it), the degree of control exercised over the premises by the owner, and the business's history of compliance with food legislation (as defined in the Code of Practice). On this basis, the frequency of inspection required to regulate each business is generated by computer. A high-risk premises may receive a biannual or yearly visit, while a low-risk business would be visited only once every two, three or even five years. For trading standards departments, the risk is based upon metrology, fair trading, food safety and product safety, and each premise is rated as high, medium or low in each category. Any problem discovered during an inspection could mean a follow-up visit; the discovery of non-compliance with any of the food regulations would most likely mean that a business would subsequently be accorded a higher hazard rating. Within the study areas, EHOs and TSOs have used the hazard analysis approach since 1994. The establishment of the system required that for many premises, the EHOs had to input information based upon their own informed judgement. All the 'superleague' retailers in the study areas were accorded the same low-moderate hazard rating, on account of the very high confidence in management and despite the very high turnover of trade.

The second component of hazard analysis is that the business itself must show that it has identified the safety hazards in its own operation and has taken steps to control them. Essentially, what this means is that the operator must identify points in operation where the hazard may occur, and then decide which points are critical to control to ensure consumer safety. These are the critical control points (CCPs) which need to be monitored and remedial action taken if safe limits are exceeded. As the Chief EHO in Stoneford commented:

> I mean, the big change that's now been worked through with the regulations is, operators of food businesses themselves, having to do their own risk assessment. And, that is such a big change, with a major sort of shift of emphasis. They've got to look at their systems, they've got to understand where the problem is, and are likely to be, and to put in place controls and monitor them and so on. Well, we have clear guidance that we're to be – certainly for the first twelve months – not to even think about taking formal action for that, because people have got to be helped. They've got to adapt to that.

Equally important as the frequency of inspection, however, is the manner of its undertaking by the enforcement officer. This has also been revised under this new system. The nature of management and the technological sophistication of the 'superleague' retailers means that inspections in their supermarkets generally involve little more than the review of systems of risk control; in comparison, lower tier establishments are perceived to require a more traditional inspection of shopfloor hygiene standards. Hazard analysis should lead to a reduction in the discretionary element of public-interest regulation, and thus one of its sources of variability through space and time.

As we have demonstrated, these varying relationships with the national and local state are the result of differing systems of internal and external regulation that operate at the level of the individual food retail outlet. Corporate retailers can fulfil their food safety responsibilities with government because of the nature of their management and hazard control systems at the local level, which ensure uniformity and consistency in food quality; the diversity and variability of in-house practice amongst the non-corporate food retailers, in addition to the lack of inter-firm co-ordination, means that such relationships are impossible. This situation now dictates the role of the local-level food law enforcement official. A system of food regulation has therefore developed that can be differentially applied to corporate and non-corporate retailers, because of their inherent differences in the operation of management and hazard control. As a senior EHO in the London borough commented:

> the hazard rating system … applies equally to all businesses big and small. But personally when it comes to doing the inspections, when you're doing the big premises, you don't need to spend too much time,

and you do feel as if you are wasting your time if you're going round picking up the odd bits and pieces ... Because the moment you look at it, they know very well what's gone wrong, somebody hasn't done something in the last 10 minutes and it's sorted.

He continued:

when one goes round it's really just ... a paper exercise ... with the management and you're just seeing how their systems are operating, and you're asking pertinent and relevant questions about the time they're checking temperature controls of frozen food coming in for instance, how long it stays out there before it goes in their storage freezers and records they've got for temperature recordings ... one goes through a lot of motions and checking up on questions ... but you can't ask these questions at the local kebab shop. It's meaningless ...

The differentiation between local level public and private regulation is further illustrated by comments from the officer involved in these site inspections:

The major firms are not bad, it must be said. In fact they tend to be ahead of us on new legislation. Well, they're in there when the new legislation is made, aren't they? ... it's rather like – along the lines of self-regulation ... they've got the resources, these people (the large supermarkets) to employ – they do employ their own EHOs, they do employ their own legal advisors – and they damn well make sure that they're really up with the latest legislative changes and advise their firm accordingly in every case. They do that quite well, I reckon ... And with these new industry guidelines – well, if they can produce their own guidelines, well we don't have to make up the guidelines any more if they can do it themselves ... very much it's a two-tiered system.

The promotion of both codes of practice and of hazard analysis allows EHOs to differentiate, in their regulatory activity, between the 'superleague' and other food retailers. Crucially, it represents an acknowledgement by government, and a willingness of the 'superleague' of retailers themselves, to in large part engage in self-regulation. As a senior official at the Department of Health commented:

Yes, I think the large retailers are very much capable of regulating themselves ... this means that the role of EHOs are reduced in their case ... I think to some extent we have now introduced an element of that in our code of practice. One of the things that we did, to help with uniform enforcement under the new Act [the Food Hygiene Directive] is produce ... a system of rating premises according to risk, which helps them to dictate the frequency of inspection. One of the weighting factors is confidence in management of business. So in other words if a business is, as one would hope a large supermarket is, very well managed, and has high standards and so on, that would effect the frequency of inspection. And there has been discussion ... as to whether, if a business is, let us say, paying for its own consultants to advise it on hygiene, and standards, as the big supermarkets do – providing it's following that advice – you could ask the question, well is there a need for the EHOs to go round so frequently? ... if a food business has got a highly organised quality assurance and hazard analysis system in place, it may be that the inspector doesn't need to go much further than that ... as we see it, it is a major change to emphasis and approach ... our evidence is that the worst of the overzealous enforcement problems are over.

In other words, the firm engages in HACCP and the regulators in risk assessments. Ideally, firms should be adhering to the relevant code of practice. The promotion of both codes of practice and of hazard analysis allows EHOs to differentiate, in their regulatory activity, between the 'superleague' and other food retailers. It is also a key step in retailer-led moves towards private-interest regulation.

Conclusions

As a result of the growing influence of the food retail sector, and a period of contestation between policy construction and implementation, a revised system of food regulation has emerged to reflect the relationship between corporate and non-corporate enterprise and the national and local state. In practice, EHOs now increasingly adopt a bifurcated approach to the regulation of food retailing. They largely retain their traditional approach, mixed with some rudimentary hazard analysis, to the bulk of independent traders. It is a style of regulation that is firmly within the public-interest mode. For the major retailers, however, the persuasive and insistent approaches of EHOs are being supplanted by an auditing approach. Corporate retailers, in contrast to the small independents, operate their own systems across all their stores to ensure uniformity and consistency in food quality. Here, the EHO in effect operates as an external guarantor of the retailers' internal quality control procedures (see Table 9.1).

It is not only possible to locate the work of EHOs and TSOs around an increasingly separated public/private dimension, but also to divorce the nexus between the public and the private with regard to the relations between the different segments of food retailing and the state, and with regard to their internal operating systems. In many cases, these processes of change are not new. What is significant is the coalescence of regulatory strategies which permits greater self-regulation, with the ability and willingness of the major food retailers to ensure that their suppliers conform to predetermined standards of food quality. The result is that a public system of regulation that has defined geographical boundaries becomes increasingly irrelevant to the strategies of the major retailers and the food that they sell nationally.

Table 9.1 A simplified model of public–private regulation within food retail stores

Private regulated (corporate)	Public regulated (non corporate)
Relation with the state	
Links with local state through Home Authority	No links, or minimal links (BIGA) with national state
Links with national state through Head Office and BRC	Links with local state geographically defined
Highly professionalised management (in-house EHOs/TSOs)	Multifunctional management
Internal operating system	
Uniformity of hazard control systems	Idiosyncratic hazard control
National level coordination of food safety and quality	Variable system
Role of local level regulators	
Provide extraneous check on quality control	Integral to ensuring base line food quality standards

10 The local regulatory interface

Enforcement practice on the ground

Introduction

In Chapter 9, we considered the recent evolution of the food regulatory system. As we saw, EHOs and TSOs are now largely engaged in proactive regulatory activity, based upon the routine site inspections of premises selling food within their local food authorities. This is organised and enacted upon according to a system of hazard analysis that is compiled within the local food authority but guided by nationwide principles laid down (in the case of EHOs) by the Ministry of Agriculture, Fisheries and Food. Here we explore the basis and nature of the persistence of local-level variation in food regulation.

The key questions that we seek to answer here are: how do EHOs and TSOs discharge their duties *in situ* with food retailers? How do they practice food legislation in their professional roles? How is this related to the regulatory behaviour of food retail outlets? Answers to these questions begin to concentrate on three areas of analytical focus. The first of these is to understand the operating context of regulators or the internal environment in which they operate and their scope for discretionary activity. It is these often highly localised conditions that help to structure the regulating behaviour of EHOs and TSOs. Thus, our second concern is with routine inspection, and with the hazard analysis by which this activity is regulated and the risk that it seeks to ameliorate. How is hazard assessed and by what means is it controlled? Third, how does the assessment of hazard vary across the tiers of retailing, regarding both public enforcement and private activity? In other words, we further analyse food regulation along the vertical axis of food retailing that runs from superstore to corner store. Through this exploration, we are able to advance the macro model of public–private regulation presented in the earlier analysis with a micro-focus linking to the local delivery of food hygiene policy.

The chapter begins with an exploration of local operating conditions for food regulators. We are not seeking to provide a comprehensive account of variation within local government, but rather to explore the range of operating conditions and the sources of variation that we came across in our study of food authorities. We wish to show how these variations can facilitate or constrain the discretionary regulatory activities of EHOs and TSOs. The chapter then discusses the types of general regulatory activity that have previously been identified in studies of the professional behaviour of EHOs (Hutter 1988; 1989). From this base, we suggest that additional *modus operandi* have evolved as a consequence of food retail change and the drive towards regulatory nationalisation that was outlined in Chapter 9. We then go on to present, across the hierarchy of retailing, the public–private interface of regulatory behaviour. The chapter concludes with a consideration of the nationalisation of food regulation around the principle of hazard, and the negotiation by public regulators of the private interest regulatory behaviour of the 'superleague' retailers. While general patterns of enforcement behaviour can be identified for the different tiers of retailing, it is clear that other localised variables interfere with the legislative attempts to provide nationwide conformity.

As the previous chapter explains, an overriding theme of the 1980s and early 1990s was the movement towards the corporate retailer-led nationalisation of the food regulatory system, to ensure the even enforcement of food legislation across the country. This mirrored the corporate retailer developments of establishing national supply and distribution systems, and as we saw in Chapter 8, their role as powerful agents in the spatial restructuring of our study areas. Indeed, our depiction of the modes of regulatory behaviour has shown that hazard analysis is similarly applied in each of the local food authorities according to legislative requirements; however, at an even finer level of analysis, within these commonalities lie subtle differences in enforcement practice, according to the specificities of places and people. This is the inherently variable spatial aspect of regulatory activity, or the *horizontal* axis of food control, and it depicts those variables that problematise the notion of national uniformity of both regulatory method and outcome.

As we have seen in Chapter 8, the nature of local retail geography clearly influences the manner of regulation (given that certain types of enforcement behaviour are associated with particular types of food retail establishments), and this itself is reflected in the study areas. The organization of the departments that we studied and the slight variations in the duties of individual enforcement officers within them also matters; more important, perhaps, is the departmental 'culture'. Related to this are two main factors: first, local politics, which are manifest through the behaviour of local councillors and magistrates; and second, those personality traits attributed to particular enforcement officers, which we will term *professional manner*. This is perhaps the most difficult to define, but as we will illustrate, certain officers have personal histories in which particular types of enforcement behaviour are (or are perceived to be) more dominant than others. As we shall show, it is officers' perceptions of these factors which influence how they regulate. These horizontal variables work to further elaborate the model of regulatory behaviour presented in Chapters 7 and 9 in both regulatory practice and outcome. They also show how the imposition of a national system of regulation is always constrained by local and micro variability.

The geography of local food regulation

The nature of place: the rural/urban divide and retail geography

Hutter (1988), in her study of EHOs, argued that proximity is an important explanation of the approaches adopted by professional regulators. The degree to which officers are integrated into the locality they serve affects not only their personal inclination to adopt either informal or formal techniques but also influences officers' assumptions about the population they control. Thus, it is *social* rather than *physical* closeness that is the crucial factor.

Officers, partially in the more rural authorities, were keen to make a distinction between the regulatory practices of rural and urban food authorities. As one officer in the rural Stonefordshire commented:

> If you worked in an inner city area – and I used to work in Birmingham, I worked there for a long time, I think in relation to a city situation, you may be in a different field of work in a way, in that there are people, when you send your formal letters to them in certain areas, they don't do anything. Whereas when you come out into – I don't know whether you'd call it a friendlier environment or what – but you know, we find that using informal measures, we get a reaction and it's not necessary for us to actually serve notices.

The officer continued:

> I don't know whether it's something that's been built up over the years. Or whether it's because we haven't got the number or the range of premises that they've got in these larger conurbations.

The visits that were made accompanying EHOs generally support Hutter's contention that those who operate in a small and fairly close-knit community generally know the people they are regulating, and they fear that the positive outcomes of legal action may be outweighed by its negative effects, both in terms of their working relationships and their social interactions with the regulated and their families. According to Hutter (1988), officers working in these smaller environments typically assume that they are dealing with good, respectable people who are in need of education and advice. Officers in such consumption spaces may be less aggressive in their enforcement strategies because it is easier for them to affect change when necessary. Further, this will be reinforced by the configuration of food retailing itself in these areas, as Chapter 8 explained: businesses are likely to have a longer history in the community than the more rapid turnover of ownership and management in inner-city retail spaces. The comments of one officer based in Tetington were noted in the field diary:

> They don't do prosecutions (they haven't done any since 1990). He said they prefer the informal approach, and usually find that it's enough to achieve the desired results. He also said that it's to do with size. Whereas in Birmingham you might have 20 dodgy premises on your patch, here you were only likely to have one or two. He also said it was to do with permanency of the business itself. In the inner city, a lot of small food businesses open up and then are gone six months later, whereas in this area businesses tend to be much more stable.

Conversely, then, it is suggested in the evidence that those working in large conurbations may adopt a more suspicious attitude. They are less likely to be acquainted with those they regulate, do not fear to the same extent the negative consequences of legal action, and are likely to adopt a cynical and less charitable view of the regulated. Not knowing the regulated well, the location and incidence of rule-breaking may be less predictable. These factors combine to suggest more frequent recourse to formal enforcement methods. In larger and more anonymous social settings, enforcement officials are policing a population of strangers, so must judge people on more objective, legal and universalistic criteria than their (shire) counterparts, who are more likely to be able to evaluate the situation in the light of particularistic knowledge. It is important not to press this point too far, as it is mediated by other factors such as departmental culture, which we explore below. Moreover, TSOs in Newton made it clear in terms that their more rural counterparts would recognise, 'that they were encouraged to develop friendly relations with traders as long as the traders realised they still had to comply'.

The degree to which enforcement officials are integrated into the community they serve also seems to be one of the mechanisms whereby they learn of acceptable and 'normal' methods of handling deviance and transgressions within a particular social setting. We pursue at greater length the enforcement strategies of officers later in the chapter, where we distinguish between persuasive, insistent and auditing approaches.

What is happening, therefore, is that the enforcement officers are variously moderating their enforcement strategies in order to achieve satisfactory compliance. This is the impact of locale and 'consumption space' (defined by the food authority) upon officers' enforcement strategies as they attempt and are increasingly pressured to secure a natural uniformity of food quality standards. Consumption space still offers them some room for manoeuvre.

To achieve the lowest common acceptable denominator, some consumption spaces are more exacting than others upon the officers involved.

Departmental culture

Just as the nature of the external environment affects the enforcement strategies of environmental health and trading standards officers, so does the internal one. Each local authority department has its own 'culture': the 'ethos' of the department, the mission statement, the personality and drive of the management, and the individual personalities of the staff. It is difficult to separate organisational 'culture' from politics, or professional manner, as all are intertwined. However, the impact – singularly or combined – results unavoidably in some degree of unevenness in the enforcement of food legislation.

The following field diary and transcript extracts serve to illustrate the variables that influence departmental culture, as they relate to the local authorities included in the study. The intention here is not to assess the extent of impact of this variation – indeed, it is not possible to quantitatively assess – but rather, to draw attention to the variables which mitigate against the uniformity of enforcement which is sought in the 1990s.

> For instance, in Cettleham the head of the department is keen to get a Charter Mark for his office, which is taking up most of his time at the moment. He is very proud of his management systems and claimed the local government auditors had congratulated him on the way the office was managed. They tried to get a Charter Mark last year and were commended for their operation but did not receive one – but he is confident that he can get one this year. Apparently it is unusual for local government departments to be awarded a Charter Mark. There are staff manuals on inspection procedures, and procedure for emergency prohibition orders and notices etc. For this officer, management of staff and their procedures was important: we put controls in, to control what we are doing and we've challenged those controls to ensure consistency of standards. One outcome of the consistency of officer performance is a limiting of their discretionary authority, to deal with circumstances as their professionalism suggests.

Similarly, Stonefordshire TSOs department was highly managerialist in outlook:

Part of their mission statement is to be the 'best trading standards department in the country'. They gave a clear impression that this is a department that wanted to 'make a name' for itself. They employ a barrister when necessary and will go after the 'big boys' when they 'need to'. For example, they are presently prosecuting a major retailer, for selling out of date food. The department has a very aggressive sampling programme – partly to show how active they are – and show a willingness to take on issues that are to do with the profession as a whole, and not just to do with Stonefordshire. They have a much more proactive stance than merely fulfilling their statutory obligations – targetting the nutritional value of infant food and baby food for example. They've taken about 20 samples, got the analyst to do a nutritional analysis on that, and then they got a nutritionist who they linked up with at the hospital to interpret those results ... and perhaps there may be certain labelling infringements which they can investigate, or compositional infringements, but if nothing else, they may be able to get some publicity out of it. And they don't share their findings with other TSO departments on the Gold Line, (like the Newton group do) but use them to campaign nationally and through the press. Staff have a perception that the Chief TSO is keen for them, as a department, to 'make a noise' to ensure that they don't get their budget cut. This need to hold on to resources and to be seen to be useful pervades much of their work.

The Department has an ISO accreditation, one of only a few TSO departments to do so. This means that they have outside auditors in every year to check that all their systems are of a high enough quality to still be worthy of the accreditation. And because of this, they are expected to do internal audits between the different divisions for each other, to check they are following the procedures as laid out (which are documented in detail, in a large manual). The ethos of the department is to be as proactive, rather than reactive, as possible. For this reason, they do not respond to complaints immediately – unless the latter are to do with an immediate health issue. Instead, staff log complaints on the computer and then try to deal with them when they do a visit.

In Newton, as with the other food authorities, there was a much greater sense of being a follower of emergent trends. As one officer outlined: 'We're trying to set up a quality assurance system here, under BS 5750 and we can guarantee a consistent level of advice and enforcement of legislation'. A junior officer in Shington went further than any other, though, when, without prompting, he made a point of emphasising the discrepancies in enforcement practices between individual officers:

> He said that he thought that different officers regularly have widely differing opinions of what constitutes risk, and of what is a 'good' or 'bad' premises. He thought that two officers from the same council could walk into the same kitchen, and then give widely different accounts of what they had seen in terms of food hygiene practices.

While such variation in practices may be a sign of a 'weak' profession, to which quality management systems can provide a welcome consistency of service, such systems will nevertheless be seen by some officers as constraining their traditional professional authority.

Local politics

Local 'politics' affect food enforcement departments in several ways. First, political decisions at the local authority level determine how much of the local authority resource 'cake' is apportioned to food enforcement. Second, local councillors can and, as we shall see, do affect the perception of regulators of prosecution policy, and also the direction of departmental sampling programmes. Finally, the perceptions of the attitude of local magistrates also affect prosecution policies, as the extracts below from the field diary and transcripts illustrate.

First of all, there is differential impact of resources:

> To be perfectly honest, because of resources, what we tend to do, and I should imagine most local authorities are like this – is that you go so far and then you stop, and you simply can't continue because of lack of resources.

However, the officer quoted here was keen to point out that the department was not short of resources:

> Well, we've been in a peculiar situation, really. Because when this new legislation came in, in 1990, we applied for new resources – for new technicians – and we got what we asked for, which knocked us over, basically. Because the government made money available in 1990 but it wasn't ring fenced, so most local authorities said, well, ok, the government's said that extra money's been allocated to the food safety enforcement function, but in actual fact our revenue support grant has been cut, so where is this money? And a lot of local authorities said well – they couldn't identify it. The government said, well it's there, you

know, if we hadn't provided this money then you would have got even less, but that didn't wash with a lot of local authorities. But we were in a situation where we managed to convince our members that this money existed and they provided us with more technician power. So we've never really, since 1990, been particularly short of staff.

Budgets were, however, under increasing threat. As one officer explained:

One target was sampling budgets which had been quite a large budget. It's now half of what it used to be about four years ago. It might not be sufficient to do all the sampling we should be doing, which is a worry frankly. It was up to councillors to confirm the sum of money, but it is down to the departments to actually decide their need to put the money into there. But, at the moment if you're, with cutbacks going on you cast round try to think oh what should we cut next. And that is one of the things that seemed to get cut.

The officer pointed out, however, that there is a 'bottom line':

if it [levels of sampling] gets too little then MAFF will come round and say look you local authority are not doing enough sampling as required under the EEC rules and that. And we're going to have to make sure it's done and do it for you and still charge you for it. If we knew we were in that situation, I don't think we are, but if we did come to that situation, the councillors would certainly make sure that the sampling budget will be reinstated. [Councillors will respond to] that sort of pressure. But it [the sampling budget] is shrinking, and it's shrinking rather rapidly. Individual samples are up to about £500 a go, so it is an expensive business, sampling.

For the more managerially minded TSOs in Stonefordshire, their departmental strategy also had a resource logic:

The other thing you have to bear in mind is that, as a local authority, we are very much about trying to demonstrate to our own elected members, and the people in this area, that we as a service are doing something that's of benefit locally and therefore should be supported. There's a political angle here, and the political angle is that *we're in competition with other local authority services* to make sure that we don't come off too badly everytime the local budget comes around. So one of the reasons that we are being more campaigning and proactive rather than just doing the mundane stuff (as I would call them, I know it's important but nevertheless mundane) of responding to complaints and enquiries, is that we feel that's the best way and a better way of demonstrating a. that there's a need for work to be done in this area, and b. that resources should be available each year, demand and duties, demand goes up each year, duties are going up each year, and resources are being cut in real terms. We've done a bit better this year, better than some of the other departments in the authority, in so much as we've had a cash increase in our budget, but in real terms, because you've got to meet pay and prices, inflation and so on, in real terms it's a reduction in what you've actually got to spend on the operations side. In the last four years, about 100 pieces of additional legislation, 25 per cent increase in complaints and enquiries, 20 per cent reduction in resources, and a 12 per cent reduction in staff [emphasis added].

As might be expected, the senior officers in the departments are more sensitive to local political considerations than the junior staff. In some cases this can lead to quite different impressions of the attitudes of councillors to the enforcement of food hygiene legislation. There is, however, a very common attempt across food authorities, sometimes based upon very little evidence, to anticipate the actions (or reactions) of councillors and for officers to modify their behaviour accordingly. There are marked variations in the relationships between officers and councillors and the former's perceptions of the latter. At one extreme, officers believe they are left alone to carry out their work as their professional norms dictate.

The Trading Standards Department in Newton has to report to the Public Protection Committee every three months or so, and the committee asks for an explanation of what the department is doing. The Chief Officer makes a presentation, and suggests certain areas to the committee on which he would like his department to concentrate and asks for their opinion. For example, the officer may suggest that the department has a crackdown on videos, and tells the committee that the department would like to report on that. An officer from the department suggested that if a department has a shrewd Chief Officer, then they can pretty much do what they think is best. However, they are answerable to the councillors.

Similarly, in Tetington an officer claimed:

Our members – as far as food safety work is concerned – adopt a hands off approach. In fact, we only report to them about once a year. And we do an activity report, which tells the local councillors what we've been doing over the year, and we tell them about some of the enforcement actions that we've taken. We don't discuss any specific cases with them. We don't have to go to them for approval to take prosecutions or closing orders or that kind of thing. The director's got delegated power to do all of those sorts of things. So it's really just a case of – the members set the policy, they've agreed that our target inspection frequency should conform with the government inspection frequency criteria. And that sets our target for the year, and we do our best to comply with that. They [the councillors] don't really have any say as to what the budget allocation is. Because the director has a budget and the managers decide how that budget is carved up. So I would discuss with my manager, how much money we think we need for the year, and that would have to come out of the overall pot. I mean we probably don't take more than about two or three prosecutions each year at the most. Last year we didn't take any, so with such a small sample of cases, it's very difficult but I would say my perception is that magistrates are more aware of what food safety cases mean. I remember probably ten years ago when I was working in Bristol, we often used to despair, because we'd sort of take what we thought were quite serious cases to court, and people would get very small fines.

In Shington, though, there was a belief amongst officers that in the local business community there was a strong feeling of 'Shington council against small businesses' and this permeated the magistrate's court. One officer suggested that there was a very strong Chamber of Commerce, and the result was that local courts were overly sympathetic to businesses:

We notify the councillors if there is anything of interest ... in real terms we have delegated authority to the chief environmental health officer, to do whatever, and they have an interest, but you know, there isn't that much to tell them. I mean if you do prosecutions and that sort of thing, yes, but we've even got delegated powers to take action on food matters anyway. I mean, we have had times in the past when we had to go to council to get permission to prosecute. And it has changed now. And we've got delegated powers to the borough environmental health officer and the chief environmental health officer, together with the solicitor who advises to actually take action against anybody. I mean what we would do if we did have a problem say, for instance, the person who we were prosecuting was a councillor, then we'd go to committee on that. It's got to be really sensitive, in other words, for us to need to but sometimes, you see, if the council don't sit or something like that, it can be a problem, can't it? And it's really a hearing, in front of a hearing, as well. Because once you start going to council, it gets put in the press.

For the most part, officers seem to be able to modify their behaviours so that they are congruent with those of councillors without too much difficulty. In Tetington, however, there was a real sense of frustration, where officers stated that they have a definite non-prosecution policy, and that the councillors have a heavy influence in this.

Policies of non-prosecution or prosecution as a last resort are common, as is illustrated below. In the food authorities we studied, the rapprochement between officers and councillors of the former's style of regulation was such that there was little need for councillors to act in an interventionist manner.

Professional manner

Individual attitudes towards enforcement undoubtedly influence the regulatory behaviour of officers. The norm is politeness, understanding, often empathy: but when faced with a legislative breach, some are 'famous' for responding fiercely, as the following examples illustrate:

> S. said that they all had different operating styles. Ja. is known to be very thorough, and to chat for hours with the people she visits. Apparently she is always late back. Jn. was described as a scatter-brain. S. described herself as being somewhere in between. She said that each of them have their own style, and that J. was famous for being extraordinarily thorough. This was obviously a good thing, but Jn. was always complaining that she took twice as long as anyone else to do an inspection.
>
> I got the impression that she (P.) is seen as a bit of a 'tough' enforcer – for instance, she was the only EHO in the country, apparently, who in one year sent a letter to the head office of a major corporate retailer, to inform them of a problem that she had found in one of their local stores, and asking them to respond immediately. In conversation she admitted that half the reason that she had done it was because no one else ever did these days.
>
> On the way back to the office H. told me how much she was irritated when they knew that people [food retailers] were endangering people's health but were unable to prosecute them. In one high profile local case of food poisoning the laboratory to whom the sample had been sent had not been able to absolutely guarantee that their evidence would hold up in court.

The above extracts illustrate that there are strong social pressures on officers to conform to the dominant operating pattern of their department. Departmental procedures and unwritten codes of behaviour all come into play to codify officers' enforcement strategies. The result is that some departments have been able to resist the shifts in national policy for implementation that we described in Chapter 9. Other departments are more sensitive to changes in the external policy environment and officers modify their behaviour accordingly. There is a final group of departments which have been to some extent anticipating national policy, and again, like the first group, these have been able to adopt much greater continuity in their enforcement patterns. It is these variable operating contexts which provide the backcloth against which general patterns of enforcement can be identified. It is to these that we now turn.

Exploring the public–private interface

Patterns of enforcement: creating compliance

Hutter (1986, 1988) argued that the enforcement of environmental health law is generally non-punitive, in that it does not simply refer to legal action. Rather, there is a wide array of informal enforcement techniques used by officials. Hutter's work was concerned with the whole range of environmental health activity rather than just food law enforcement, but none the less she identified common enforcement practices amongst officers. Securing compliance was the main objective of these professionals, both through the remedy of existing problems and, above all, the prevention of others. And the preferred methods to achieve these ends were co-operative and conciliatory. Thus where compliance was less than complete, and there was good reason for it being incomplete, persuasion, negotiation and education were the primary enforcement methods. Accordingly, compliance was not necessarily regarded by enforcers as being immediately achievable; rather, it could be seen as a longer-term aim. The use of formal legal methods, especially prosecution, were regarded as a last resort, something to be avoided unless all other measures had failed to secure compliance.

Moreover, Hutter's work in a variety of Environmental Health Departments distinguishes two distinct strategies within the broad approach of compliance. They are referred to as *persuasive* and *insistent*. Both share the common objective of securing compliance, as opposed to effecting retribution, but they differ in the stringency of the means to these ends. The persuasive approach epitomises the accommodative approach typical of regulatory agencies, and the range of tactics favoured by those adhering to such a strategy are informal. Officials educate, persuade, coax and cajole offenders into complying with the law. They explain what the law demands and the reasons for legislative requirements. They discuss how improvements can best be attained. Patience, understanding and support underpin the whole strategy, which is regarded as an open-ended and long-term venture.

The insistent strategy is less benevolent and flexible than the persuasive approach. There are fairly clearly defined limits to the tolerance of officials adhering to this strategy. They are not prepared to spend a long time patiently cajoling offenders into compliance, and they expect a fairly prompt response to their requests. When this is not forthcoming, these officials will automatically increase the pressure to comply.

Since Hutter's analysis, as we have seen, the food retailing (see Chapters 3 and 6) and regulatory (see Chapter 7) world has changed. To accommodate both the new regulatory and retailing environment in which they work, new regulatory strategies have also emerged which are sensitive to particular types of retailing establishments. As the discussion here will illustrate, persuasive and insistent strategies are still central in the work of the officers studied across the different food authorities. We show how these strategies are nuanced by officers as they judge retail outlets. In addition, a distinct third regulatory strategy has emerged in which officers engage in the *auditing* of a firm's hygiene management systems. This strategy is targeted towards the corporate retailers, but is bringing other firms into its orbit.

The evolution of regulatory complexity at the micro level

Persuasive and insistent strategies are generally now employed with the bottom tier of food retailers. The rise of the corporate retailers, together with the swinging pendulum of the food regulatory environment of the early 1990s (outlined in Chapter 9), has encouraged the introduction of new techniques for application with the top tier. As the corporates have evolved, persuasive and insistent strategies have been found to be inappropriate for firms that are privately regulated. In the main, enforcement officials accept that the corporate retailers are 'ahead' of public regulation, and so when they undertake an inspection, they are acting as benign external auditors of privately regulated systems. This is based on a new trust and expectation: the public regulators trust the retailers to fulfil their part of the private interest equation. However, commensurate with this are very high expectations; and, on the occasions when enforcement officers may 'catch' the corporates underperforming in a regulatory sense, they adopt an authoritarian mode and demand immediate response. The presumption is that the corporate retailers 'should know better' and that the officers' 'trust' in them has been abused. This differentiation is summarised in Table 10.1 and explored more thoroughly below.

Analysing hazard and controlling critical points

The work of food law enforcement officers, and food regulation on the part of retailers is now, as Chapter 7 described, based around the identification of hazard. In individual retail outlets, hazard analysis and control is a central constituent of the due diligence clause; in council offices, the frequency of regulatory visits to each of the food premises under the jurisdiction of the 'food authority' is also determined by a hazard assessment. In the latter case, the following transcript from an interview with a senior EHO is illuminating:

Well, when we do an inspection of a premises we carry out a risk assessment. It's basically the same assessment that's described in the Code of Practice number nine. The Department of Health issue Codes of Practice on enforcement, I think there's about seventeen of them altogether. But Code of Practice number nine deals with food hygiene inspection, which basically lays out the kind of protocol if you like, for doing food hygiene inspections. And it tells local authorities that they must have a risk assessment system, and it sets out I suppose what you would call a 'model' risk assessment. Most local authorities follow it – they don't have to follow it, but at least must have a system which is similar. We've probably been implementing it since about 1992 or '93. So when we do an inspection of a premises, these are the categories that we look at. We look at the potential hazards, which is basically what type of food the premises prepare. And they're given a score according to how risky the food is. You know, low risk foods, if it's just a shop, that handles packaged foods, and that kind of thing, they'll get a very low score. But if they prepare high risk foods, they'll get a high score. If they actually are a producer of high risk foods, a manufacturer, then they'll get a higher score. We also have a weighting for different types of processing, which are considered to be fairly hazardous – cook chill, and that sort of thing. And also the size of the business. The more consumers who are supplied by the company, then the higher risk they are. And we also look at the compliance. How well they do. Not necessarily the history, but what we see when we go. So we'd look at how well the business is complying with the legislation. There is some guidance in the Code of Practice as to what all these things mean. But I suppose it is fairly arbitrary in a sense. But there is an indication of how well they comply with food hygiene and safety requirements, the structural standard of the building, and then finally, the confidence in management. Whether they've got documented control systems, whether they carry out food hygiene training, you know, doing risk assessments and that sort of thing. There are weightings according to the size of the business, so obviously a very large organisation like Sainsburys would probably be considered to be an intermediate, whereas a corner shop would be supplying very few people. So that weights it accordingly. Similarly, you'd expect Sainsburys to be of a very high structural standard, and so it would get a very low score for its structure. All that information is put into the file, and then one of our clerical officers will add that onto a data base. And – according to the score – that determines the risk category of the premises. There are six risk categories, A to F. And category A gets an inspection every six months, and category Fs every five years. So based on the number that goes into the system, that will then determine when we revisit.

TSOs also utilise a hazard rating system on which to prioritise their regulatory activity; this however, is devised by each food authority rather than according to national principles.

Hazard analysis on the part of food retailers themselves is based upon a different set of procedures. In the generic, it is based upon Hazard Analysis Critical Control Point, better known as HACCP, as discussed earlier. The hazard analysis portion of HACCP was intended to identify sensitive ingredients and sensitive areas in the processing of ingredients or food where critical points must be monitored to assure product safety. Critical control points are those areas in the food production chain, from raw materials to finished product, where the loss of control could result in an unacceptable food safety risk. The Hygiene of Foodstuffs Directive (Council Directive 93/43/EEC) specifies that food business operations must identify and control any step in their process which is critical to ensuring food safety using the HACCP system.

The corporate retailers have been introducing HACCP into their safety systems since the 1980s. However, the expectations of the regulators of the hazard assessment utilised by the independent retailers are far less sophisticated. Instead of a complicated procedure, independent retailers may be required to provide evidence of a check list that proves that they have given thought to the areas of greatest risk to food safety in their premises.

Regulator meets retailer

It is clear that the different tiers of retailing elicit different types of regulatory response from enforcement officers that involve a reconfiguration of expectation, trust and support on the part of the regulators. Although corporate retailers are subject to more regulation *per se*, enforcement officers are expected to be 'gentle' in their application of the law. This is evidenced in their dealings with all kinds of retail establishments. But so too is the need to strictly uphold safety standards; a gentle and supportive role can, when necessary, be replaced by authoritarian zeal. The different dimensions that can be involved in the enforcement strategies of regulators in their relationships are outlined in Table 10.1.

As we have already indicated, officers are likely to pursue persuasive and insistent strategies with the lower tiers of food retailers. Auditing of the management systems of the corporate retailers is much more likely to occur. We do, however, provide an example of a bakery below to show that the auditing approach is gaining ground amongst firms outside of the corporate retailers. The different strategies that officers adopt are linked to the levels of trust that they have of a firm, and this largely shapes the actions that they take. For example, the persuasive approach is linked to high levels of trust in

Table 10.1 Strategies for enforcement in action

Officer action	Enforcement strategy			Level of trust
	Persuasive	Insistent	Auditing	High
Supportive	√		√	⬆
Steering	√	√		⬍
Authoritarian		√	√	Low

the food standards of a firm and will result in officers being supportive of efforts to improve standards further. Any lapse in standards would lead to the officer intervening more directly and steering the firm so as to raise its performance. Levels of trust in the corporate retailers are also high, but here, if an infringement were to be discovered it would lead to more authoritarian forms of action.

Below, we now concentrate upon the key modes of behaviour and patterns of enforcement identified in the fieldwork, and elucidate some of the key dimensions in Table 10.1. This is accomplished largely through the composition of transcribed material, which has been chosen on the basis of its ability to demonstrate the key patterns experienced. The evidence is selected from both field diaries and transcribed interviews. Much of this description of enforcement documents three-way interaction between the enforcer, the retailer and the researcher.

Independent retailers: the persuasive approach

We begin with an exploration of the work of regulators on the interface with the lower tier of retailing to assess how they ensure compliance with regulation. As we have already argued, their favoured strategy is one of persuasion. Within this approach, officers generally like to be supportive of businesses they trust to be safe, and know are well-intentioned. This trust is based on previous knowledge and often long-term relationships between the regulator and regulated.

When an officer feels no additional action is needed by the business, when it is doing everything it can to eliminate risks to food safety, and probably doing more than is necessary to satisfy regulatory compliance, praise for the efforts of the business is likely. The following example of a visit to a fish and chip shop and bakery illustrate this case well.

The second example is of a bakery shop in the High Street.

The premises was really very small – just a kitchen with fridges and freezers, another room where the potato peeling machine was and then a toilet. G. went to check this first – everything was in good order – and then we had a quick look at the back yard. There was an old chest freezer out there, unused of course – but a hazard because of children climbing into it and suffocating. (Later on when G. mentioned this to him, he said that he had been phoning the council to ask them to come and take it away for ages, but with no response; G. promised to get it sorted out for him.) Then we went back into the potato cutting room. Of course this is a messy operation – with water on the floor, but apart from this the room was completely tiled and obviously thoroughly cleaned at the end of every day. The only problem was that the pipe to the hand basin in the room was broken (and later he said that this was a temporary thing; G. said of course she realised that problems occurred and was definitely confident in his explanation). There was a raised, dry area, for storing the bags of potatoes. We then went into the kitchen and opened up the freezer doors to reveal a highly ordered fridge. At this point Mr F joined us, to explain his rotation system. He showed us which fridge was used for fresh fish and which freezer was used for frozen; he explained his rotation system for his curry sauces. There was only one complaint – an opened tin of pineapple rings (should never be stored in the tin) and he apologised and said that 'we'd caught him out on that one'. Then he pointed to a gleaming silver bucket in the corner and said that that was what he used to make his batter in. There were two spotless chopping boards: one for fish and one for salad. And in another freezer there were some ready made pies, neatly ordered and labelled. He explained their rotation system. At this point G. interrupted him. She said that she always liked to give credit when credit was due, and she wanted him to know that his whole set up was absolutely excellent. She said that she was thoroughly impressed with the way that he was running his business.

The second example is of a bakery shop in the High Street.

The bakery shop sold cream cakes and ready made sandwiches, bread and non-perishables. R. said that the shop was operating well over and above the legislative requirements. There were two middle aged women working in the shop which was part of a small chain of local bakers, owned by local people. They said that the manager comes round regularly to make sure that the place is spotless and all the risk prevention strategies are being followed. They were instructed to throw away sandwiches once they had been in the refrigerated display cabinet for more than four hours. R. said that this wasn't necessary from a hygiene point of view. But they said if the manager discovered sandwiches that were more than four hours old they would be 'in serious trouble'. We were there about 50 minutes, and R. told them what a good job they were doing before we left.

In the three examples below, we show how EHOs and TSOs act to ensure compliance.

First stop today was at a small declining butchers near Newton town centre. This was a family business that had lasted for several generations but was now struggling. In its documented history, at least, it had never poisoned anybody: S. appeared to be very sympathetic to this man. At the end of the inspection he was asked only to replace some worn away tiling behind the counter and some of the rusting bars in the meat hanger (which was the original ice room from 1918 and was under the floor. The rubbish removal system was fine, and the butcher had got a pest control contract with someone. S. mentioned the 'hazard assessment' to him, but said that he wouldn't expect him to write anything down on paper because he obviously knew what the procedure was and would be the person who carried it out.

Ja. and I only had time for one visit before we had to get back up to the offices. It was a kebab shop in the centre of town. The place really was in very good order and I was surprised at how very thorough Ja. was. The floors, work surfaces, fridges and freezers all got checked. There were some minor requests for changes 1. the television cable was a health and safety hazard and so had to be taped up 2. the edges of the work surfaces needed resealing 3. the freezer needed a new seal 4. there was no hot water available for the hand sink. There had also been a complaint of noise from the shop's extractor fans. Ja. was very sympathetic. She explained to the owner that she could understand that it would be very difficult for him to remedy the problem, but that she had to inform him that there had been a complaint. She suggested that he shouldn't take action yet, but wait to see if there was a further complaint in the future. She also spent at least 10 minutes explaining to him about the hazard assessment approach to controlling risks on his premises, and gave him a copy of each of her leaflets. She asked him about whether he had had any food hygiene training and he said that he had.

On a visit with a TSO we went to a really nice old fashioned store with very beautiful tiling and huge old counters. Even some old scales. One of the scales in his shop was slightly out but as soon as it was pointed out to him, he was very obliging and said that he'd get it fixed immediately (whenever E. finds anything like that wrong he always pays the premises another visit shortly afterwards to see if they've sorted the problem out). E. then explained the new rules on metrication and the shopkeeper listened very carefully. He seemed worried about the conversion tables but said he'd take action by the end of the summer.

Directive to supportive

Below, we illustrate the case of a dairy in Stonefordshire that had previously presented a potentially serious health risk, but where, after a period of direction, a more supportive relationship is once again being established.

Last year, there was a series of bad results from the samples of milk, as far as bacteriological quality goes, from a big local dairy and so S. was doing a test every week to monitor the problem and make sure that it wasn't getting any worse. After things improved they went back to six weekly visits, because it takes up a whole morning for her and so they really only want to maintain the minimum necessary surveillance. S. didn't have to introduce herself as they all know her there. S. had a quick chat with the laboratory technician who is responsible for on-site testing of laboratory quality. It was a very friendly chat. S. then selected six cartons from which to take milk quality samples. S. told me that the dairy firm had recently made some alterations to the premises, which had been advised by the environmental health department and she showed me some of the new machinery that had been installed. She said that the physical quality of the premises wasn't all that good, because the turnover of business had

> expanded so much that the capacity of the premises itself wasn't really adequate. However, it was a very important local employer and they [the EHOs] understood the problems that the company faced in keeping the premises up to scratch, and so they tried to help them in every way they could, because they were basically a good business.

Not only does this example illustrate the nuances of enforcement, but it also brings out well the way in which regulators will try to understand the problems that business face in dealing with competition and the value of a firm to a local community. One of the problems that regulators will increasingly face is that their perceptions of a firm are based upon their training and experience. Willingness to act in a flexible manner where appropriate may become more difficult as quality management systems and hazard analysis increasingly begin to structure behaviour and responses.

Independents: auditing

As we have already argued, most independent food retailers are subject to a traditional regulatory relationship with EHOs. Below, however, we illustrate the cases of a bakers and a restaurant where hazard analysis is to the fore. The bakers is subject to the private-interest regulation associated with their critical supply chain relationships as well as the regulatory activity of EHOs. More specifically, they are drawn into an auditing of their hygiene systems by both their customers and the regulators.

> L's is a baker with about six of its own shops in the local area (including one in Tetington). It employs about 100 people. As well as supplying their own shops, they also make deliveries to other stores and cafes. G. had set this up as a routine inspection, but he was also intent on showing me how cooperative EHOs can be, and how they want to help businesses comply with the legislation. He is obviously of the opinion that L's is a top rate business. When we got there we had to sign a form, before we were allowed onto the premises, saying that we had not travelled overseas in the last six months or suffered a stomach upset. (I later found out that an external private auditor had insisted on this stipulation, and L's had to do this to keep a contract with one of the people that they supply.) G. and the manager engaged in a routine inspection. They went through all the manuals for about an hour, where the HACCP etc. were documented. The manager said that the amount

of paperwork had increased exponentially over the last few years. He said that his business had improved as a result of all the new regulations that were in place, but that he had had to destroy a rain forest to do it. He often asked G. for his opinion on the way in which they had done things. G. assured him that he was operating at a standard higher than the legislation necessitated. G. wanted to look at the complaints book as well. The manager said to me that he didn't hide anything from G. He wanted him to see it all and understand what problems he had running such a business. M. said that the big problem he had was not with the EHO department from the council – he genuinely did give me the impression that these people he counted as his allies. G. said that he wanted M. to be completely honest with him, and he accepted that there would also be food hygiene problems in this sort of business. M. said that the biggest problem was from the external auditors that other companies employed to audit him, in order that they could prove their due diligence if they took supplies of bread and cakes from him. He said it was the 'large small' business that was really suffering from all of this. The very large companies had exacting standards – as was right – but they were big enough to be able to afford to put all the systems in place. The problem was that they expected all these standards to be replicated exactly in small companies as much. He lost an NHS contract for supplying bread rolls recently, even though he'd been supplying them for years, when a private auditor came round. Because he didn't have health records for his bakery staff, and they hadn't had stool samples, or something like that, when they started work with him. He's now implemented this system, but he lost the contract anyway.

We put on our hair nets to do the bakery inspection and even though we'd brought our own whites with us, they wanted us to use theirs, which had a protective cover over the buttons. When we went into the bakery, G. reminded me that it was a messy business, and not to be surprised by flour etc. that was all over the place. We inspected all the machinery, and G. wanted to know what the cleaning rotas were, and we inspected the storage cupboards. Only one problem was found – some of the wooden shelves upon which dried baking ingredients were stored were splintering, and there was a box of dried fruit without its lid on, so that fragments of wood could get inside. G. used a kind of familiar cajoling to make his point – 'what the bloody hell do you call this, come on man, you wouldn't have a leg to stand on if someone brought a loaf to me with a piece of wood in it'. His attitude throughout, I felt, implied that he knew that they were really doing their best to run a good business, but that didn't mean that they shouldn't take him absolutely seriously. The company was up to date on all the legislative requirements – that was clear – but they wanted reassurance from G. that they were fulfilling things correctly. And G. wanted to satisfy himself that the systems that they had on paper were actually working in practice. It is a clear example of the auditing procedures spreading to the more advanced companies that are locked into supermarket supply chains. In total we were there about two and a half hours.

Today we did one routine visit to a new restaurant. J. wanted to check HACCP documents with the manageress. We were there for about 45 minutes. We looked at the kitchens and the store rooms but mostly J. was interested in the management systems: temperature controls and records, staff training in food hygiene issues and health and safety issues. J. also asked about the kind of menus that they had drawn up for the summer (upon entering any restaurant, EHOs look at the menu in order to assess the potential risk posed by the types of food provided e.g. prawn cocktail is high risk, reheated chicken and pasta is high risk, chips are low risk). J. talked at length about the new emphasis on this control approach. I was aware (not for the first time) that busy catering staff probably think the EHO sounds ridiculous at times, so much of it is common sense. I got the impression that she was thinking – doesn't he realise that I've got a busy business to run?

Corporate retailers

For many corporate food retailers, the procedures of hazard analysis are now widespread if not always well practised. In accordance, EHOs have had to make changes to the ways in which they regulate. Below, we describe and illustrate good practice in relation to hazard management; finally, we illustrate how hazard control can lead to a more authoritarian response from the regulators.

Not surprisingly, officers have adapted in different ways to the changing emphasis of food regulation. As one officer explained:

she had not yet quite become comfortable with the idea that she was now encouraged to do inspections without actually inspecting, so to speak; that is, the idea being that instead of actually looking at the walls and floors, she should look at the management systems and written procedures. This is illustrative of the general change in approach since the early 1990s; instead of being obsessed with the detail of the tiles and formica, the risk assessment approach puts most of the emphasis on systems of operation and critical control points.

According to S., this really is the opposite of how things used to be.

Meanwhile, another interviewee described in the field diary was more hopeful:

All we're doing really is looking at the systems that we've got and testing them, and seeing if what they say they do actually happens in practice. So, and I think ultimately, hopefully, that's what will happen when people finally get to grips with hazard analysis, that's what our inspections will focus on. We'll be looking – talking to them about the systems that they've got in place to control food safety ... and then just checking, and auditing them if you like, to see whether they actually work and whether they're doing it in practice. Because that's what often happens, you know ... people have a nice glossy folder on the shelf of the office, and they can show you all the flow charts and everything. And then you go into the kitchen, and talk to people in the kitchen about what really happens, and it bears no relationship to the system.

A third officer had sufficient experience of the auditing approach to express some cynicism at his own role:

The codes of practice applies equally to all businesses, big and small ... But personally when it comes to doing the inspections, when you're doing the big premises, you don't need to spend too much time, and you do feel as if you are wasting your time if you're going round picking up the odd bits and pieces, the odd thing that's gone down, a bit of dirt here or there. Because the moment you look at it, they've seen you look at it, they know very well what's gone wrong, somebody hasn't done something in the last ten minutes and it's sorted. When one goes round the big supermarkets it's really just sort of, a paper exercise, it shouldn't be but that's the way it goes. With the management you're going through the paper work and you're just seeing how their systems are operating, and you're asking pertinent, and, and relevant questions about the time they're checking temperature controls of frozen food coming in for instance, how long it stays out there before it goes in their storage freezers and records they've got for temperature recordings. One goes through a lot of motions and checking up on questions.

Below we report how this officer carried out a visit at a major retailer, in which he tries to mix together an audit and a brief inspection.

Before we announced ourselves S. did a quick check around the super-market. To begin with S. checked that the shelves were clean and tidy. He then checked the sandwiches section, because he was interested in the dates of all the takeaway food. The manager had rearranged his morning schedule to spend time with us (we were there at least four hours); he had also arranged that one of the company's Safety Officers be there. To begin with we talked about the shop's control systems. There was a system which provided safety information on all aspects of the shop. This has replaced the manuals that each shop used to have, and details all the procedural information that they need. S. was very interested in this and later on we looked at it on the computer in the manager's office. There was another system that checked all the fridges and freezers. If any of them went below the required tempera-ture for a certain period of time then an alarm bell sounded. In order to turn off the alarm, a member of staff had to log in with their pass word. Therefore somebody was forced to take responsibility for it. There was another system which required the store to notify head office of any complaints it had received regarding food products or any problems it had found itself. This allows head office to work out if there is a problem with a product nationwide, so it can isolate it and withdraw the product (and charge the manufacturer for compensation, including payment for person-hours lost). We also discussed the withdrawal systems. If head office does identify a problem then stores are notified immediately to withdraw that product. All stores also get continual weather reports, so that they may plan accordingly to which foods might be in greatest demand. After this we put on our whites and went on a tour of the meat cutting room (soon to be defunct), the big storage fridges and the unpacking areas. S.'s only complaint was some damaged tiling, and the manager said that the builders were meant to be sorting that out. We saw the crusher that turns all the rubbish into bales, and we watched a delivery going on where all the refrigerator stuff is marked with red tape and gets priority in the unloading bay. The manager told me there was a computerised inventory system so that they knew everything, down to the last can of soup, that was on the premises.

Hutter's (1988) study predates developments in refashioning and regulation that we have documented; and the simple binary conception of persuasive and insistent regulatory modes is inappropriate for the contemporary situation. The research found the local-level food enforcers operating, with the 'superleague' retailers, largely as supportive auditors. Here, their visits operate as external checks on the privately regulated internal HACCP systems of each superstore. This mode of operation on the part of the regulators is based upon a very high degree of trust, and expectation that the retailers are operating well above the standards expected by law.

This public regulator–private interest relationship requires and assumes that the corporates are constantly 'overperforming' in all aspects of food safety hazard mitigation. Given this, on the occasion that a public enforcer discovers a breach in this trust, the reaction is beyond insistent, and is directly authoritarian. The binary persuasive or insistent strategies do not apply. The opposite side of the regulatory equation is that the local-level food law enforcers have much higher expectations of the corporate retailers. The equivalent standards of compliance, that with an independent store may elicit a persuasive or insistent response, would lead to immediate severe and possibly punitive action with the corporate retailer sector. As one TSO said, without prompting, he would be much more likely to prosecute a major high street name for a violation than he would a market trader, because he would expect a high street chain store to know better.

However, between the 'superleague' and independent retailers are a group of corporate food retailers (supermarkets and fast-food chains) that are neither employing the sophisticated HACCPs of the 'superleague' or the basic identification of hazards that is expected of the independents. They illustrate a form of regulatory 'no man's land'. As we will see, instances of 'underperformance' elicit an insistent rather than persuasive response. They are judged more fiercely and expected to respond more quickly than the bottom tier of food retailers.

S. had only had two complaints about the pizza company documented on his system before we made the visit, which he didn't think was very much. We began the inspection in the kitchen. There were broken tiles around the front counter and the floor was dirty. What was also surprising was that the manager was very happy to point out to us the things that were wrong. She said that she was fed up with contacting head office about the things that needed doing, but never getting any response. The table upon which all the pizza toppings were placed was directly next to the very hot oven so that the employees do not have to walk around with prepared pizza in their arms, but it means that it is very hard to keep the toppings cool. S. found his first serious problem with this table. It had on it about twelve different containers of toppings, each indented into the table which was supposed to act like a fridge. S. stuck his thermometer into one of the containers and it only went down to about 12 degrees – not the five degrees that we would have expected. The manageress said that it had been faulty for a while and that she had complained to head office but so far they had failed to respond. On the wall all around were hygiene instructions to staff and instructions on how to make each of the different types of pizza, but each of the containers were uncovered, and that annoyed S.. There was a piece of salami in one of the containers full of tomato sauce. S. then checked all the cold storage – both the temperatures of the fridges and the freezers, and the dates on the food products. He was pleased with most of the fridges, but not with the way that some of the food was stored in the freezer. Some of the desserts – such as pies – that had slices missing from them were stuck on top of the freezer without any wrapping, waiting for the next customer to order some. He suggested that they be kept in containers. There was an old dough mixing machine also standing in the kitchen that hasn't been used for a long time. S. suggested that they get rid of it because it was a hazard and the manager said that she had been requesting head office to organise it for a long time. We then went to check the storerooms at the back of the premises and the rubbish disposal. S. was appalled by the way the rubbish was stacked in black plastic bags in a tiny space outside the back door of the shop, blocking the fire exit. The manageress was too.

We also checked the staff room and the toilets. S. thought that the staff room was a health hazard for employees because there was no ventilation. S. spent a long time with the manageress in her office, trying to find the log book for staff and customer accidents, and the information on pest control. When she eventually found the former, S.

discovered that there hadn't been an entry for about three years and wasn't very impressed that they weren't bothering to document them anymore. When the pest control documents were found they were quite damning. The pest control company had discovered lots of problems and had written several entries saying that urgent attention was needed. These things hadn't been carried out.

While we were waiting for all of this, S. added all his information into the hand-held computer and came out with the risk evaluation of the store. This was higher than it had been previously estimated, and so the store will be getting more frequent visits, as it was deemed to be of moderate, rather than low risk. S. also said that he would be getting in touch with the head office of the firm, using the home authority rule, i.e. he will get in touch with the local authority where head office is, and tell them about his worries and what needs to be done in the Newton store. It was clear that in working out the risk evaluation for this store S. was judging it more stringently because it was a big chain store. For some of the questions used for the risk evaluation equation, he was tapping in 'poor' into the handheld – but he said he would have judged it as 'moderate' if it had been a small store.

Insistent and authoritarian behaviour

Of course, enforcement behaviour is not always openly supportive or clearly insistent; sometimes officers can be a little cryptic in their handling of the circumstances that greet them on a visit to a premises. Again, this is influenced by prior knowledge of the business's compliance record. It is common for officers to disguise their insistence as support. Officers may feel that the easiest route to compliance is to allow the business to feel that its problems are understood, and its efforts to achieve compliance (through identifying and controlling hazards) are recognised and supported. In the following examples, the officer involved is clearly unhappy with the situation; however, she offers to come back and give the business time to 'clean up its act', rather than use punitive measures.

We called in on a pub in the High Street which made meals. As soon as we got there, the proprietor began the sort of defence that I have heard before – explaining that he knew things were wrong, and that it was only the week before that he had had to sack the chefs because of their appalling standards and had taken over the running of the kitchen himself. But, of course, had not yet had time to get everything back up to standard. We weren't there very long; H. looked in the kitchen and so 'oh dear, they did make a mess didn't they'. Apparently about a year before, a lot of money had been spent on improving the kitchen, as H. had previously witnessed. After this, the proprietor had 'let' out the kitchen to a couple, who were then responsible for running the food side of the business. He now said that it was them who had made the place so dirty – it was clear that a lot of the dirt was 'old', i.e. ingrained. H. didn't stay long; she just did a cursory inspection of the toilets, after she had looked in the fridges and asked him a few questions about how long he was keeping the very large amounts of rice we watched him preparing, and how it was reheated before it was served. She said, 'well of course you realise that it can't be run like this, but if you've only been in for a week or so I understand'. She said she would be back this time next week. Once we got outside she said that she suspected that he'd probably been running the kitchen himself for quite some time, and had just not bothered cleaning it up adequately. She thought a week was sufficient to get it up to scratch if he knew that she was coming back, and wouldn't be prepared to accept such poor standards.

For EHOs or TSOs to go further and adopt more stringent action is rare. It does, however, occur. When hazard is apparent and serious, officers are willing and able to act. What also emerges from these two examples is that the response of the regulated is taken into account when deciding action. In both cases, there is a sense that those running the business are not playing the regulatory game by breaking trust or faith.

Example one:

I asked G. about the department's prosecution procedures. She said that they had recently had a successful prosecution, but that it was the first one since 1966. It was her inspection that had led to the prosecution, and a food seizure, which is very unusual. During a routine visit to a shop she found a whole stack of 'blown' cans of raspberries. She told them that they had to be removed. Samples were sent to the public analysis and found to be contaminated with yeast – enough to so that eating the raspberries could make you feel pretty ill. When she went back to the shop to tell them about the findings a week later, she discovered that they were still selling all the tins. So she went back to the office and got another EHO to come with her, and they seized all the food. She said the blasé attitude of the management was quite astonishing: they genuinely didn't seem very concerned about being prosecuted.

Example two:

> H. showed me some photographs of the kitchens of a business that they're trying to prosecute at the moment. Two people had taken a temporary lease over the Christmas period in a pub to serve Christmas lunches. H. said that they had obviously gone into this as a way to make a huge amount of money very quickly and in this respect, apparently, they had done very well. When H. went to inspect the premises, they were serving lunch to hundreds of old people, but the conditions in the kitchen were such that H. wanted to prosecute immediately. H. said it is a very frustrating case. The couple were probably making £5,000 a night, yet would be unlikely to be fined more than £300. They would use the temporary nature of their set-up at that particular premises as an excuse. They are now trading in another place, and therefore publicity that will be attracted to the case will not even affect them.

Conclusions

We have been able to show that, in general, public regulators at the local level of enforcement operate differentially along a vertical axis of retailing. A new form of 'benign' behaviour has developed to accommodate the corporate retailers, one that differs from the supportive, persuasive or insistent patterns of enforcement used with the bottom tier. 'Supportive' strategies are also 'new', and are a strategy devised by EHOs and TSOs to help 'good' local businesses within the changing retailing environment. This is modified by different expectations of what these different types of retailers can do to control their hazards, but also because of the generally low risk hazard that it is assumed they represent. However, because the expectations of the sophistication of corporate HACCP is so high, public enforcers will actually react immediately in an 'authoritarian' manner with corporate retailers in the event of breakdown of that system. Corporate food retailers who fall outside of the top three 'superleague' occupy something of a regulatory 'no man's land'; insistent rather than persuasive strategies are used here to enforce higher standards than those expected of the 'bottom tier' of independents.

This typology of local level regulatory behaviour is based upon the nationalisation of food regulation which presupposes that legislation is evenly applied across space, but tacitly differentially applied between the tiers of retailing. Hence, despite considerable differences in the geography of retail space in the two study sites, as far as food enforcement is concerned, the main axis of variation is hierarchical rather than spatial. The nationalisation of food regulation imposed (and discussed in Chapter 9) in the 1990s has allowed for hierarchical strategies of enforcement at the local level associated more with type of store/outlet rather than type of place. The micro-empirical and analytical approach adopted here has exposed at the interactive level of enforcer–retailer the variable and discretionary attempts made by enforcers to carry through food legislation to enforcement. A more variable set of enforcement strategies is the result, and these tend to operate consistently across the different responsibilities of TSOs and EHOs.

Particular enforcement strategies are clearly related to the expectation of a business that an officer has from its previous history of compliance, and the 'hazard' presented by the business at the time of the visit itself. The stance adopted by the enforcer will therefore constantly vary from business to business.

11 Conclusions

Retailing, regulation and consumption

Introduction

The various literatures in the subject areas of retail and economic geography, rural sociology and food policy studies have begun to focus upon the significance of retailers in the provision of foods in advanced economies. This study, of the British case in the 1990s, has tried to contribute to this work by integrating corporate retailing developments with both the regulation and consumption of foods. We have attempted to show that the regulation and provision of food is conditioned by a particular combination of public and private interests in Britain. In particular, in all three sections of the volume we have sought to integrate the active role of corporate retailers, government and consumer organisations and their representatives. It is out of the interplay of these three spheres that food provision is regulated, different types of foods are put on offer and the 'competitive space' for corporate retail capital is created.

Such an approach, based on detailed empirical work at European, national and local levels, holds implications for our broader understanding of modern state intervention systems and the nature of corporate retail competition. Government agencies find it more difficult to control the food sector given the rise of the corporate retailers and political pressures to restrain from too much intervention. They are, as we shall reiterate below, forced into compromises of action which attempt to represent the consumer interest. Corporate retailers, on the other hand, however sophisticated their systems of supply and distribution and shop management, need to increasingly engage in what we have termed 'the politics of market maintenance'. This means that they have to develop relationships of cooperation and competition both with each other and with state agencies. As we argued in Chapter 3, they have to play a full part in a range of regulatory domains in order for their strategic economic objectives to be met and possibly maintained. It is this sociopolitical economy which has therefore conditioned the nature of food regulation in the UK in the 1990s, and which has played such a pervasive role in restructuring not only the nature of food purchase and consumption, but the very nature of towns and cities (as we see in Chapter 8), and the ways that food choices and outlets are locally regulated (Chapters 9–10).

The triumvirate of concerns in the book (of food retailing, regulation and consumption) has been focused upon developing a more integrated regulatory analysis of food interests in Britain. As Clark (1992: 616) states in calling for more attention to be given to the regulatory state:

> the administrative manner, style, and logic by which the state regulates society in general, and the economic landscape in particular ... involves ... moving away from the theory of the state to the nature of administration per se (and) can be interpreted as the next step in a process of refinement and finer-grained research.

Taking such observations seriously, we have applied these to a sector which has undergone considerable change and turbulence as the 1990s have progressed. It is a sector in which, despite a growing negative image on the part of consumers (food sales) and considerable confusion on the part of the state as to its most effective role, a major corporate sector – the retailers – has managed to increase its role, rates of return and overall supply-chain power. The conceptual-level analysis in Part I, as well as the national and local empirical analyses (in Parts II–III) document, at these different levels, how these conditions have been achieved.

In undertaking this analysis, we have also paid particular attention to two other major issues in the understanding of the geography of regulation. These concern the questions of regulatory level and the role and significance of space. These two dimensions are linked together, particularly in Parts II and III of the book. In terms of regulatory level, it has been important for us to consider the ways in which combinations of policy and corporate retail strategy transcend EU, national and local levels of regulation, as well as to understand the interactions between the evolving formation and trajectory of policy and its micro or local implementation. These dimensions come together in Part III, as we consider the ways in which local regulators, whether planners (Chapter 8) or food enforcement officers (Chapters 9–10) implement and interact with the retail sectors at the local and national levels. We have shown that emerging out of these micro-regulatory relationships is a new model of regulation which treats different sectors of the food retailing trade differently. The model is co-evolutionary, and tends to shape the types of food choices provided according to the type of outlet.

In terms of space, as Part III outlines, it is clear that in a very material way both changing food policy (see Chapters 2 and 4) and corporate retail strategy (Chapters 3 and 6) have influenced local retailing and regulatory spaces. However, this is not as clearcut as we might imagine. Retailers have been intent upon trying to reduce the significance of local circumstances in the provision and regulation of their supply chains; and they have tried to influence government policy in ways which centralise and 'de-space' food supply and food consumption problems. Government, on the other hand, as our emerging regulatory model in Chapters 7 and 10 indicates, while responding to these corporate pressures, is still left responsible for the localised regulation of quite markedly different food spaces. It still needs to run an effective local authority-based food enforcement system. Our emerging regulatory models therefore contain different implications and definitions of regulatory space.

We have also seen, in the use of the concept of 'competitive space' (Chapters 3 and 6) how retail food markets have to be created out of different spatial implications of (1) retail store location competition, (2) competition in supply chains between the retailers and the upstream suppliers, and (3) the increasingly intense intra-sectoral competition between the different tiers of retailing. These competitive spaces are made out of the interactions between retailing, state activity and their variable representations of the consumer interest. They are moreover, as our analysis has shown, anything but static in the food sector. They have to be created week by week, and this suggests a growing fluidity of the evolving food regulatory system in the UK. We have shown that a key dimension of these simultaneously competitive and regulatory relationships concerns the nature, codification and articulation of food quality criteria (Chapter 6). Contestations as well as coalitions around the constitution of food quality pervade all of the relationships between retailers, the state and consumers. Retailers for their part need to constantly assemble legitimate conceptions of food quality upon which they can trade and compete (as well as keep state regulators happy). This book has indeed been about the story of how retailers have struggled to do this in the 1990s, producing what seem to be legitimate conceptions of food choice. (This means 'freedoms to', as opposed to 'freedoms from'; see Chapter 4.)

This process is, of course, a never-ending story. By the late 1990s (not least with BSE, *E. coli* and GMOs) we see new challenges and contestations in the regulation of food in the UK. One of the key questions becomes how the retailer-led food governance system developed analytically here will be sustained in the near future. Will it be subject to major problems of consumer consciousness? How will the leading retailers continue to construct legitimate conceptions of the consumer interest? Is the system of food governance in the UK sustainable economically or socially? These are questions beyond the scope of this particular book. In conclusion, however, we will attempt to outline two key prospective aspects of these relationships. First, we will consider in summary the issue of the social and political location of the corporate retailers and their attempts to maintain their 'competitive spaces'. Second, we will consider this in relation to the continued re-regulation of food in the late 1990s, with particular reference to the broader regulatory significance of the protracted BSE crisis, and the Labour government's attempts to appease consumers' anxiety by establishing a new Food Standards Agency.

Both of these developments tend to reconfirm, we argue, the established nature of a retailer-led food governance system, one which is typified by private-interest regulation. The analytical models established in the book, therefore, are seen to demand further specification as their relevance in understanding the trajectory of food consumption rights and citizenship becomes even more embedded into British consumer culture.

Sustaining retail power: maintaining competitive space

By the late 1990s, despite considerable problems emanating from the growing consumer consciousness about food and health and about the domination of food markets by the retailers (see OFT 1997), the corporate food retailers have managed to tighten their grip on the provision of food choices. While this may have become more difficult, due to the intensity of competition, the tightening of planning siting policies and the need to innovate in providing different types of quality supply chains, the consensus is that these hurdles have largely been overcome. Indeed, as Wrigley (1998) and Langston et al. (1997) indicate, any tendency towards 'saturation' has been offset with increases in the efficiencies of opening smaller stores, town-centre convenience stores (like the Tesco Metro centres) and the diversification of inventories (particularly petrol). More emphasis has been placed upon focusing on the 'recurrent consumer', ensuring through such devices as 'loyalty' cards that consumers return regularly to the same stores.

However, concentrating on these more recent 'surface' strategies in consumption relations tends to hide some of the progress made by the corporate retailers in further embedding their dual role with the state. In more profound terms, corporate retailers and state agencies have continued to see the advantages of mutual purpose in progressing the consumption orientation of 1990s Britain. We have outlined in earlier chapters some of the general trends here, namely, the deregulated state of the 1980s and 1990s being prepared to allow a vibrant corporate and consumption-oriented sector to take responsibility for a major section of food regulation and the social provision of foods.

This socio-political strategy has been largely mutually reinforcing. The nation state of the 1980s and 1990s cannot control the development of consumerism in society. It needs retailers to project and to implement consumption. In doing this, retailers – as multidimensional complexes of consumption knowledges – become important social barometers of consumer reaction and activity. They become more important than recognised consumer groups (see Chapter 5), for instance, in assessing what the consumer will tolerate. If knowledges about foods are to be disseminated, it is to the retailers the government departments now tend to turn to undertake this. This has most recently been seen with the reactions of supermarkets to irradiation and genetically modified foods. Scientists and MAFF officials recognise that the retailers perform the role of key consumer testers as to the legitimacy of these new food innovations, even after the MAFF and DoH regulatory committees are satisfied.

By the late 1990s, then, and despite a growing history of state 'limits' in the quality regulation of foods, corporate retailers had emerged as the 'clean' providers of ever-increasing food choices, and had done so with the government's blessing. The development of a Food Standards Agency – an attempt by the Labour government to be seen to clean up the food system and to re-legitimate the public-sector regulatory structures concerning food (see below) – is largely accepting of the corporate retailers' commanding position. The agency will concentrate upon providing a more consistent and detailed regulatory structure for the public regulatory system of providing baseline standards of food safety. This leaves retailers with the freedom to develop and to innovate in their own quality hierarchies (see Marsden *et al.* 1998). Retailers have been keen to gain uniform treatment of their outlets from the multitude of local authorities in which they are located. They have largely achieved this through the adoption of the Home Authority Principle (see Chapter 7) and the development of more consistent applications of hazard analysis, conducted by the local authority EHOs. The ability of such retailers to manage the quality of foods through their supply chains (and indeed to add value to these foods at the consumer end of the chain) has also been enhanced by the passage of the new regulations. As a result, new relationships have been forged between public and private systems of regulation (see Chapters 7 and 9). It is increasingly apparent that the public system of regulation that has focused on implementation in geographical (local authority) boundaries is increasingly irrelevant to the diverse sourcing strategies and regulatory abilities of the retailers and the foods they sell nationally.

For the large number of smaller, independent retail outlets the situation is, however, quite different. They too must now make some effort (at significant cost and from an extremely low practical base) to identify and control hazards within their operations. This is further reinforced by the recommendations of the Pennington Report (1997) into the *E. coli* outbreak, which embodies the need to 'accelerate implementation of HACCP for high risk premises' (1997: 32). It is here that the state regulation remains important for maintaining food standards. Corporate retailers have regularly pressed a somewhat increasingly beleaguered government to implement tighter food control systems more generally, and particularly onto the heterogeneous independent retail sector (see House of Commons Agriculture Committee 1998). Armed with both an apparent customer legitimacy and consistent systems of food supply chain hazard control, they have presented models to government about how this should be achieved and implemented.

What we see, therefore, is a highly differentiated style of food regulation emerging in the late 1990s, particularly in the degree of regulatory and consumer scope it will be able to encompass. At one end we have a public regulated system based upon a spatial approach associated with local authority enforcement. This is being pressured to improve its standards of enforcement. The principle is to provide baseline food safety standards and to enforce these, particularly with regard to the independent retailer sector. However, under the intense competitive conditions in the selling of foods, the corporate retailers offer more elaborate quality food choices through their supply chain management systems. The multi-outlet retailers' definitions of food quality exceed the current regulatory requirements. Given such diversity, simply to call for a uniform raising of standards is neither sufficient nor of economic benefit for the corporate retailers. They can outplay the independent sector in terms of quality and then demonstrate the necessity of their practices to government for enforcement of the public system of regulation. The independent retailers are increasingly feeling the costs and stigma of being located at the sharp end of baseline food regulation, while the corporates occupy a more lucrative territory or 'competitive space', one which is located at the apex of quality consumer choice.

The rather ironic feature of this yawning dichotomy operating in the regulation and competitive spacing of food provision is that as government has been forced to act to more effectively ensure baseline standards, and to interfere particularly in the meat supply chains as a result of the BSE crisis, the more this seems to be reinforcing the divisions in quality food provision between the corporates themselves and between them and the independents. Moreover, it is the latter, along with the primary producers and the food manufacturers, who are having to bear considerable costs of new, government-inspired regulations. This point has recently been behind the current calling in of the corporate retailers to the Office of Fair Trading, following the livestock farm crisis of late 1997–early 1998, and the contention that both the burdens of collapsing markets in red meat and the extra cost of regulation are being passed back down the chain, from the supermarket to the producer and the processor (see Welsh Office 1998).

These continually embedded advantages instilled in the corporate retailing system, and the retailer-led food governance system more generally, continue to provide major competitive advantages over the independent sectors and, moreover, rates of overall return on capital which allow the corporates to continue to restructure and control their own food supply chains along their own sets of priorities. By the late 1990s, most of the leading retailers were heavily investing in centralised vacuum packing and processing of meats and in the more careful enrolment of selected groups of farmers from which to source. This is undertaken on the basis that it is in 'the consumer interest'. It means, more than ever, that the 'construction of the consumer interest' is increasingly delegated to a private-interest type of regulation: a form of retailer-led food governance. The social provision of quality food choices becomes even more associated with the type of retail outlet from which foods are purchased, with a hierarchy of choices extending from the handful of dominant multiples through to the discounter retailers and the independents. This is the socially constructed pyramid of food choices.

It is important to recognise, however, that such a set of conditions is far from stable or uncontested. As the 1990s have progressed, the conditions have emerged out of socio-political as well as economic factors. For the state – when things do go drastically wrong with the food supply system – still has to pick up its traditional legitimatory function, however much it has devolved its responsibilities for affecting quality food supply. The new regulatory conditions surrounding food consumption – under what we call retailer-led governance – still requires the state to balance responsibility for the public interest. This balance, however, now has to be based upon restricting traditional-style government regulation and facilitating and extending the 'sponsorship role' of government to the private sector.

We can see from these conceptualisations of the nature of food consumption regulation (as we argued in Chapter 6) that there are clear relationships between regulatory culture and consumer culture, and these will influence the social and spatial provision of foods. In these relationships, the significance of particular private-interest models of regulation, particularly associated with retailer-led food governance, becomes a feature of the overall shift of emphasis in regulation towards consumption as opposed to production (see Marsden and Wrigley 1995). Indeed, we can posit a set of mutually reinforcing relationships in the cultural political economy of the modern British state, whereby both a private-interest model of food regulation and a highly individualised quality consumer culture tend to co-evolve; giving a platform for the further economic sustenance of the corporate retail sector. In this sense, and as much of this book has demonstrated, economic power is derived and feeds off these social and political constructions. It is an outcome which feeds off particular social and political formations.

Regarding the effect of these socio-political conjunctions upon actual food consumption relations, it is worth reminding ourselves here of the Deleuzian argument (1992) which suggests that the 'trick' of modern consumer-oriented societies is to project consumerism as revolutionary new freedom, a seduction, a pleasure as well as a 'freedom to', at the same time as creating a new form of control. This form of control, ostensibly through the 'market', convinces and generates compliance by projecting the benefit of self-interest and choice options. To quote David Clark (1998: 27):

> Consumption is a system that permits capitalism to project itself in the image of individual freedom whilst simultaneously ensuring the conditions of its own reproduction. The freedom it promotes amounts however, to a freedom of an extremely limited kind: 'such freedom of expression in no way subjects the system, or its political organisation, to control by those whose lives it determines' (Bauman 1988: 88). Indeed, consumerism bears a thoroughly duplicitous character, presenting itself as the paragon of freedom whilst representing its highly specific reduction to one particular area of life. It simply defines away other questions of freedom – in particular the freedom to influence or determine the nature of the social system in the first place.

These constructions of 'control' are not only linked to the end user–consumer relationship, but they extend outwards to potentially limit the possibilities of alternative retailing provision: for instance, a vibrant independent sector, shorter, more ecologically defined (e.g. organic) supply chains, or a more progressive consumer politics. They extend upwards into connected, retailer-led supply chains, affecting the market entry and economic survival of the producer and the manufacturer (both at home and abroad). They limit choices as well as create them, and they do so in ways which uphold a particular set of power relations in the social provision of consumer choices. It seems to do so, as Bauman (1996) suggests, in ways which enrol the losers into the plot as much as the winners, enlisting discontent and contributing to a consumer-based social order by maintaining a highly unequal set of market and competitive relationships.

We should stress, however, that these sets of conditions have to be continually constructed. In terms of the prevailing commercial, regulatory and consumer cultures developed, it needs to be recognised that these conditions are highly volatile, in part because they are at root dependent upon the maintenance of individualised as opposed to collectivised consumer behaviour. The contradiction here needs continual management by both retailers and the state in order for some level of continuity and consensus in the commercial and consumption cultures to be maintained.

It is important therefore to recognise the inbuilt fragility of the current food regulatory order, and to recognise that to understand how it operates and evolves may indeed help to suggest alternative frameworks of food provision. Without celebrating the current conditions, it is possible to trace what we regard as a profound socially restructuring force. This challenges our understanding of the significance of power in food consumption. We need to move far beyond the assumptions of a dichotomy between consumer sovereignty *vis-à-vis* structures of provisions approaches (vertical as opposed to horizontal conceptions) (see Glennie and Thrift 1992; Fine and Leopold 1993). We need to resist the bland attribution of power, and to do so in ways which expose the more subtle ways in which it expresses itself. This indeed is the challenge for food studies; to not only transgress but also to deepen an understanding of power relations in providing and consuming. Focusing on the shifting nature of regulation (as it encompasses in the British case, retailers, government and consumption) is one means of achieving this.

The model of British food regulation and provision detailed in the book therefore is anything but static. As we face the end of the decade we can outline how it is still coping with the accommodation of the continuing crisis of confidence in food consumption (for example, the protracted BSE issue) and in governments' repeated attempts to re-establish some level of authority over the provision of foods. Neither pressure, however, tends to alter the current model of food governance established in the 1990s; but it does keep it active.

Food re-regulation and the continuing food crisis

This book has documented the nature of food regulation and retailing in Britain. At a time when government has faced almost unprecedented challenges in dealing with concerns from all those interested in food standards, we have investigated the issue by combining primary and secondary sources with different scales of analysis, from the local to the European. Below we wish to return to key themes in the book: the nature of food regulation, the changing face of food retailing and the marginal contribution of food consumer groups to the policy process. We wish to illustrate the centrality of these themes to food policy issues through two cases: first, a long-running but deepening food crisis, namely bovine spongiform encephalopathy (BSE), and second, a proposed institutional reform, that of the Food Standards Agency. In their very different ways, these two examples highlight the shifting nature of public and private regulation. In the case of BSE, government has found its legitimacy and credibility further undermined, and the private sector has often been to the fore in trying to resolve questions of public safety. While the Food Standards Agency is an attempt to restore governmental authority to the regulation of food, it is undermined because it fails to take sufficient attention of the realities of private-interest regulation. Finally, we wish to reflect the way in which our analysis of regulation of food can inform our understanding of governance and the maintenance of corporate retail capital.

Bovine spongiform encephalopathy

What makes BSE a particularly interesting case study for inclusion here is that as the crisis has developed and its implications for human health become more serious, so the roles of the key actors and the basis for their actions has become clearer. Below we elucidate the positions of each of the key actors in relation to BSE: government, food retailers and manufacturers and consumer groups.

Government

From the outset, the government found itself in a difficult position with regard to BSE. Like other food safety issues, BSE raises questions about the government's regulatory stance. The promotion of individual private rights appears anachronistic when consumers demand of government that it provides them, again, with a collective freedom from worries about unsafe food. BSE has also highlighted two other issues for the government. On the one side, the timing of the emergence of the disease counted against MAFF. The late 1980s were marked by a series of food scares, notably salmonella in chickens and eggs, in which MAFF had been widely perceived to side with the producer against the consumer. It thus entered the debate enjoying only limited public confidence. On the other side, matters were compounded by a belief that MAFF, and the government more generally, was partly responsible for the situation. This was because its deregulatory stance meant that MAFF failed to act on a proposal of the last Labour government to tighten licensing procedures for the recycling of animal protein into feed. Instead, according to an internal MAFF consultation paper, ministers felt it would be better for the industry to 'determine how best to produce a high quality product and that the role of government should be restricted to prescribing a standard for the product and to enforcing observance of that standard' (quoted in the *Guardian*, 23 March 1996). As a result, the proposals of the previous Labour government were dropped and a more self-regulatory regime adopted.

MAFF's actions have throughout been closely linked to a particular conception of the contribution of science to decision making. Decisions are to be made on 'good science', that is, where there is scientific certainty which civil servants and ministers can rely on. MAFF waits for evidence to prove the need for certain types of action. The problem in relation to BSE has been that science has been working in conditions of uncertainty, controversy and public anxieties that have demanded more than the pronouncements of conventional science. The public and other European countries, particularly Germany, have shown a much greater willingness to embrace the principle of precaution as a basis for action. With the advent of a Labour government once more, a more precautionary line has now been advanced in Britain.

An early indication of the Conservative government's approach was its setting up of expert working parties into different aspects of BSE (e.g. the Tyrell and Southwood committees). Equally revealing was the way the Conservative-dominated House of Commons Agriculture Committee, in its enquiry into BSE, treated Professor Richard Lacey, a leading exponent on the potential risks of BSE. (There have been allegations that at the time and subsequently MAFF provided much information to discredit Lacey; see the *Guardian* 23 March 1996.) Lacey argued for the need to adopt a more precautionary stance, but was subjected to aggressive questioning. In its report the Committee pointedly noted:

> that not all scientists carry equal authority ... Professor Lacey, in particular, showed a tendency to extrapolate sensational conclusions from incomplete evidence in order to publicise his long-standing concerns about food safety. The result was a mixture of science and science-fiction – a quite unsuitable basis for public policy.
>
> (House of Commons Agriculture Committee 1990: para. 74)

Thus, the government has engaged in incremental policy shifts as knowledge has grown. Measures in relation to BSE have included, in July 1988, a ban on beef offal from animal feed; in August 1988, the slaughtering of all cattle suspected of having BSE; and, in October 1989, the total ban on human consumption of certain cattle offal such as brain, spinal cord, thymus, spleen and tonsils. What is significant here is that beef offal was banned from animal feed before that of the human food chain because of the belief at the time of the government's scientific advisers that animal diseases could not jump species to humans.

At the highest levels of government, ministers consistently argued that: 'There is currently no scientific evidence that BSE can be transmitted to humans or that eating beef causes CJD [Creutzfeldt–Jacob disease]' (John Major speaking in December 1995, quoted in the *Guardian*, 21 March 1996). There was considerable furore, therefore, when the finding emerged from the Spongiform Encephalopathy Advisory Committee (SEAC) in March 1996 that a new form of CJD could be linked to BSE. The Committee concluded from its study of ten people who died of CJD that 'the most likely explanation at present is that these cases are linked to exposure to BSE before the introduction of the specified bovine offal ban in 1989'. When the Ministers of Health and Agriculture announced the findings to the House of Commons on 20 March 1996, a new phase in the BSE crisis emerged. Additional controls were announced: all carcasses from cattle over thirty months were to be deboned in specially licensed slaughterhouses supervised by the Meat Hygiene Service to ensure all trimmings are kept out of the food chain, and a ban on the use of all meat bonemeal in farm animal feeds was imposed. These moves were to ensure that the risk to humans of transmission of BSE was 'minimised'. In a significant shift from their earlier pronouncements, ministers no longer claimed that there were no risks attached to eating beef but rather that the risk was extremely small (*Guardian*, 21 March 1996).

The tension between different types of scientific knowledge and appropriate policy responses has also been reflected in the government's dealings with its European partners and with the Commission. Germany especially has been active in promoting a precautionary approach. This has involved going further than the British government in introducing restrictions on the sale of beef and beef products because of the potential risks to humans. The standard response of the government has been to claim that such measures are not introduced to protect public health but as a device to restrict trade. In essence, the difference between the two approaches comes down to where the burden of proof should lie. For the government, food is safe unless scientific evidence shows otherwise, while for the Germans, in the present circumstances products should be proscribed until it is certain that there is no risk of infection (*Guardian*, 5 August 1996). Not only do these differing approaches give rise to different policy responses for the eradication of BSE, but they also have variable implications for the food choices available to consumers. The precautionary approach, perhaps more than any other, is concerned to protect consumers in general from potential risks.

Once it became clear following the statements of ministers to the House of Commons that there was a strong likelihood of BSE being passed on to humans, France, Germany and Sweden acted immediately to ban the import of British beef. By the following day, all nations within the EU with the exception of Denmark and Ireland had also banned the import of British beef on health grounds. The Commission soon followed suit, making its decision binding on all member states. The decision quickly got caught up in domestic and European high politics of integration and support for agriculture, and the implications for the consumer tended to be ignored.

Nevertheless, there are three points that are worth making about the position of the consumer in relation to BSE and human health. First, as an editorial in the *Guardian* (11 March 1996) pointed out, some consumers, because of the choices they have made, are likely to be more affected than others by the government's delay in banning beef offal from human food:

> for a further unnecessary 15 months (the time between the ban on animal and human food), brains and the spinal chord of cows – the offal parts which harbour the BSE disease – were minced with other beef parts for cheaper forms of burger, pies and sausage rolls. Predictably, the poor who bought a far bigger proportion of these cheaper products, will be most at risk.

Second, British consumers were able to eat meat that other governments of Europe had banned because they regarded it as potentially unsafe. Whether this matters in practice depended on the outcome of the third point, namely, what counts as safe and who has the authority to pronounce that food is safe. At the outset of the crisis, the government argued that beef was safe. In 1990, as the then Minister of Agriculture John Gummer and his daughter ate beefburgers, he claimed: 'It's delicious. I have no worries about eating beefburgers. There is no cause for concern.' By 1996, amid fears that BSE could be transmitted to humans in the form of CJD, such categorical assurances disappeared and the possibility of a small risk of transmission was admitted. Within a matter of days, however, a small risk had become safe. The Minister of Agriculture, Douglas Hogg, sought to reassure consumers: 'the risks of eating British beef today are extremely small – or to use ordinary language: British beef is safe' (quoted in the *Guardian*, 29 March 1996). Six months later and Hogg had become even more bullish: 'The controls in place mean that British beef is absolutely safe and there is a strong case for other countries to put similar controls in place for their industries' (quoted in the *Guardian*, 17 September 1996).

Other governments saw the matter of safety rather differently. The Dutch government, for example, planned the destruction of the 64,000 British cattle in the country and advised consumers not to eat British beef (*Guardian*, 28 March 1996). Germany, meanwhile, banned the import of British beef and beef products and urged consumers to buy only certified German meat (*Guardian*, 11 July 1996). The different approaches of Germany and Britain to safety (linked to their adherence to the precautionary principle, see above) were well illustrated when it was revealed in August 1996 that BSE could be transmitted from mother to calf. The response of MAFF, based upon the advice of SEAC, was that the findings did not pose an increased risk to public health and, therefore, that no additional measures were required. In contrast, Germany felt that this was a new development and demanded an extended culling programme to eradicate risk 'to humans'.

Whatever claims the government may have made about the safety of beef, they were undermined in two ways. First, they were treated with a growing disrespect by the public. Opinion polls showed that a large majority of people did not trust the government and most blamed it for the crisis. Second, faced with a mistrustful public, major food retailers devised their own strategies to stabilise consumer confidence, and sometimes these further undermined the government's position (see below). Conveniently detached, it would seem, from the 'dirty business' of meat production and processing, yet clearly culpable for accelerating the speed of the meat supply chain, retailers were not a focus of public concern. Before interpreting the role of retailers, we briefly consider what influence consumer groups have exercised in this area.

Consumer groups

Despite the potentially enormous impacts of BSE upon consumers, consumer groups have been only 'bit' players in the BSE debate. The paradox is that while the public reacts with disgust to reports that cows have been fed on other animals, have changed their eating habits and generally distrust the government in this area, consumer groups have little input into decision making. Throughout the crisis, consumer groups have been firing from the fringes with little discernible impact; reinforcing their constructed marginality, as outlined in Chapter 5. Instead, the views of the consumer have more often been 'represented' by food retailers and, to a lesser extent, by the European Commission.

Nowhere is this better illustrated than when fears emerged that BSE could be transmitted to humans in the form of CJD. All the consumer groups could do was to highlight the dangers of eating beef and demand that the government reassure the public (*Guardian*, 26 March 1996). For example, the Consumers Association was quick to warn that those 'who want to avoid the risk of BSE have no choice but to cut out beef and beef products from their diet. There is currently an unquantifiable risk in eating beef.' They continued: 'Until we know that BSE is eliminated from all stages of the food chain, no one can guarantee that beef is safe' (quoted in the *Guardian*, 23 March 1996). The consumer, in a situation where the potential risks to health were so great, was asked to exercise restraint in consumption: do not choose beef. The Consumers Association was dismissed by Food Minister Browning, implicitly for its lack of expertise: 'Our scientific advisers are very clear that if they had felt the need to give that advice they would have done so' (quoted in the *Guardian*, 23 March 1996). Meanwhile, the National Consumer Council (NCC) also recognised that MAFF had been tardy in dealing with BSE and consumer concerns. It made two points: that there should be an independent food agency, and that 'consumers become stakeholders in the Whitehall machine' (letter from Ruth Evan, Director NCC, to the *Guardian*, 3 April 1996). All in all, there was little indication that the government felt that criticism from consumer groups merited much of a response. Their criticisms were dismissed, and neither was there any attempt to placate them by involving them in further discussion – as representatives of the consumer – on how the situation might be resolved. In this case far better placed to represent consumers, if not *en masse* then at least *their* customers, were the food retailers.

Food retailers

For the food retailers, the continued problem of BSE represented a major challenge. They needed to be able to retain their customers' confidence in the quality of their products. At a time when government has often seemed to be unsure of itself and unable to sufficiently respond to consumer demands on the safety of food, it has been the retailers who have increasingly taken the lead. Initially, this was through the setting up of a third party body, the Food Safety Advisory Centre (backed by the retailers in 1994). More recently, as the government's ability to regulate meat food standards has been questioned (see *World in Action*, 13 November 1995; *The Times*, 13 November 1995), so the retailers have been called upon to guarantee quality through their own supply chain regulation. For example, in a full-page advertisement in the national press (*Daily Telegraph*, 18 January 1996) the Meat and Livestock Commission asked: 'Can you serve British beef with confidence?' Its answer was to show pictures of the shop fronts of Tesco, Sainsbury, Asda, Safeway, Somerfield, Morrisons and Leos, and it went on to say:

> Your favourite supermarkets are usually in hot competition. On one thing, though, they're in total agreement. British beef ... Just as they have always sold it with the utmost confidence, so they continue do so ... Rest assured, your trust in their standards is trust well founded.

As the health risks of BSE have become more apparent, so the retailers have increasingly defined their own line on food safety as distinct from that of the government. The first and clearest sign was in the response to the link between BSE and CJD. Faced with the prospect of a mass consumer boycott of British beef, again, the major supermarkets initially appeared unsure of what action to take. The corporate affairs manager at Tesco summed up the feelings: 'We are monitoring what our customers want. We will get a feel for that in the next couple of days. This is a new experience for all of us' (quoted in the *Guardian*, 23 March 1996). At the outset, none of the supermarkets contemplated withdrawing British beef, but all were preparing to source from elsewhere and increase orders for alternative products, such as poultry, pork and lamb. Of the supermarkets, the Co-op was the only one that exclusively stocked British beef, and it therefore found itself in a situation similar to that of the fast food chains who also relied on domestic sources. Unlike the latter, though, the Co-op and other supermarkets were able to promote alternative meats. For the fast food chains, who were almost synonymous with beefburgers, it was a much more difficult situation. McDonald's, the market leader, was the first to respond. Full-page advertisements in the national press on 25 March 1996 announced that the firm 'will no longer be offering products made with British beef', and carried a statement from the company's president and chief executive:

> Our customers expect us to take a lead – and we have. We believe they can eat at McDonald's in confidence. We continue to have complete faith in the quality and safety of the food we sell in our restaurants. Our hamburgers only contain prime cuts of beef in which BSE has never been detected ... We believe that British beef is safe. However, we cannot ignore the fact that recent announcements have led to a growing loss of consumer confidence in British beef which has not been restored. We have always put our customers first. They trust us to provide high quality, safe food. We believe that they want us to take this action in the circumstances.

McDonald's had strong grounds for taking their action. It was reported that private polls conducted for the company showed a 60 per cent refusal to eat British beef, double the resistance to British eggs during the salmonella scare. The day after McDonald's decision, three other fast food chains – Burger King, Wimpy and Wendy's – also abandoned British beef. In their own full-page advertisement (on 29 March 1996), Burger King explained that 'our customers' lack of confidence in British beef, the related potential damage to our business and threat to our employees' livelihoods has caused us to take the decision to source beef outside the UK until confidence in British beef is fully restored.' Similarly, a Wimpy spokesman stated: 'Wimpy believes it must change to non-British beef to maintain absolute confidence in the safety of its products'. An indication of the impact of BSE on the fast food retailers came from Wimpy, who reported that: 'We have experienced a drop of 20 per cent in our beef lines, which equates to around 100,000 meals a week which have either switched to non-beef products or disappeared totally' (quoted in the *Guardian*, 1 May 1996).

The supermarkets took less dramatic action, continuing to sell British beef but promoting it heavily. For example, Sainsbury cut the price by 50 per cent and emphasised (yet again) the 'quality' of their supplies. As a spokesman for Sainsbury explained, their 'Farm Assured scheme', which guarantees meat supplies from farms to shops, had helped to secure customer belief in the safety of beef (*Independent*, 1 April 1996). Retailers also made much of improved information to consumers. There were reports that retailers were proposing a new 'kite mark' on beef to reassure customers that it was from BSE-free herds. Beef was also relabelled according to its area of origin in an effort to rebuild consumer confidence. This implied not only that Scottish beef has a cachet, but that the incidence of BSE is also much lower than in England. The Meat and Livestock Commission (MLC) (itself beleaguered on the one hand by government and continually pressured by the retailers) also launched a regular series of advertisements in which it promoted the Quality Minced Beef Mark, a sign that the beef was offal-free and from cattle less than thirty months old.

The role of the supermarkets and fast food retailers in the BSE crisis is revealing about the changing regulatory structure and the persistence of the private-interest model. First, it shows quite clearly that at least some consumers are working with different ideas of risk and safety to those of the government. The government's assurances of minimal risk being equivalent to safe food were either insufficient or disbelieved. The government's problem was that while working with current 'scientific knowledge', it could not give categorical guarantees of safety or quantify risks. As Sir David Naish, President of the NFU, pointed out in relation to the culling programme, 'science is not enough'. For these consumers, it was a different principle for action that was to the fore, one which the fast food chains also recognised: act in a precautionary manner to remove the risk. For the fast food chains, such action was almost a necessity in an effort to stabilise their market.

Second, there was a paradox to be resolved in this case. Britain has by far the highest incidence of BSE, and therefore presumably greater risks. Yet British consumers were quicker than many others to recover their confidence in domestic supplies. Sales of beef had been in long-term decline, but within a matter of weeks were back to about 85 per cent of their level before the link to BSE and CJD was made public. Opinion surveys showed that consumers do not place great trust in the government. The fast food chains had to respond to consumer fears by resourcing their supplies, but still significant quantities of British beef are being purchased from supermarkets.

Although the retailers have not emerged unscathed from the crisis, it would seem that the paradox is partly explicable because of the high levels of trust that consumers place in the quality (part of which is safety) of foods that they can purchase from the supermarkets. This in turn depends on their ability to exercise private regulation of their supply chains. In a sense, this is testament to the longevity and authority of the private-interest model. The counter-argument might be that a potential flaw in the proposed Food Standards Agency of the Labour government (proposed in 1997–8) is that it works with an outmoded model of food regulation, failing to give sufficient weight to corporate retailers' supply chain activities: in particular, their degree of trust creation with a wary public. Also, as the analysis of the research outlined in this monograph suggests, the problem for the 'Agency' is compounded by its simplistic ideas on the way in which 'street-level bureaucrats' (i.e. Environmental Health Officers and Trading Standard Officers) operate. We turn to these points below.

The Food Standards Agency: a return to public-interest regulation or increasing the capacity of retail-led food governance?

The government's recent proposals for an independent Food Standards Agency (MAFF 1998) have attracted much positive comment. They have been presented by a government apparently intent upon cleaning up the food business. The proposed Agency, which could take over the Ministry of Agriculture, Fisheries and Food's responsibilities for food standards and safety, is designed to restore public confidence in an industry that has been hard hit by a series of well-publicised food scares, to raise the quality of food that is eaten, and to advise ministers. These are ambitious tasks, and we would have some reservations as to whether the hopes vested in the Agency can be realised.

Of course, one outcome of organisational reform is a sense of action by government, and such a move on its own may well help to restore public confidence in British food and in government food policy. What it may be less successful in achieving, however, is realisable improvements in the implementation of food regulation. As much of our previous analysis outlines, positive developments in this regard require an understanding not only of formal responsibilities, or the quest for more transparent and publicly accountable structures, but also of the processes of regulation. Our research in this area suggests the need for customised and flexible rather than homogeneous and structural policy responses. This is because it is necessary to deal with variability both in retailer power of their supply chains (in the different tiers of retailing) and in the forms of regulatory practice. We explain each of these points in turn below.

Retailers and food quality

As we have seen in this book, associated with the different tiers of retailing are different notions and regulatory practices concerning food quality (Chapter 6). It is the corporate retailers who have led the way, not only in terms of innovative forms of competition and food design, but also in how to regulate food quality under increasingly complex and competitive food supply chain conditions. It is these retailers with which we have been most concerned, because of their market position and regulatory input.

The 'big three', and Marks and Spencer, are widely perceived as leading the field and have developed their own quality definitions of foodstuffs, which go well above and beyond the more limited food safety and hygiene legislation and its regulatory implementation. These firms have established numerous consumer panels and focus groups and have developed complex systems for testing consumer reactions. They have broad and hierarchical conceptions of food quality which extend beyond the physical properties of the foods themselves and include quality signals which span the process of shelf selection through the point of purchase to domestic preparation and digestion. Moreover they have, as we have seen, increasingly gained the economic and political strength to impose their conceptions of quality upon their suppliers.

The development of retailer-led food hygiene and hazard systems means that increasingly these systems become conditions of market entry for food suppliers and manufacturers. Hence, as far as the overall supply chains are concerned, it is not enough from the point of view of the retailers to supply quality foods of the right compositional standards. It is also necessary for suppliers to demonstrate that systems of quality management have been put in place (such as the Hazard Analysis and Critical Control Point system (HACCP)) as a food assurance scheme. Hence the retailers expect more and more from their suppliers in terms of the policing of food delivery as well as the type specifications of the food produced. This tends to give retailers a market advantage with customers, and demonstrates to government that the retailers are taking existing food regulation seriously (particularly the 1990 Food Safety Act and the 1996 Food Hygiene Directive; see also Pennington Group (1997)). For the customer, this means that food quality is both highly differentiated above the state-defined baselines, and that the choice of retail outlet will affect the type and the constructions of quality purchased and consumed.

Thus, in formulating the functions of any 'new government agency' we must recognise that such an agency will be working with a highly differentiated retail sector, which operates with different notions of food quality. At the peak, the multi-outlet retailers' definitions of food quality will exceed current regulatory requirements. Given such diversity, the Agency would need to adopt more imaginative regulatory responses than simply calling for a uniform raising of standards. This belies the sophisticated and innovation-led growth in private (retailer-led) governance of the food system. Any new government initiative in this regard needs to recognise that the experience of intervening in food supply and consumption will be far removed from its postwar, 'freedom from' capabilities (see Chapter 4).

Regulatory practice

These emerging developments have important implications not only in any potential structural changes in food regulation, but also in the evolution of differentiated regulatory practices which match these changes. Thus, the ability and power of the major multiple retailers to define and enforce their own food quality standards arises because of their ability to privately regulate their own supply chains. This is a novel development, and one that contrasts with the traditional (postwar) regulatory style based upon notions of the public interest. In the latter case, it is central government that sets standards through legislation and which is enforced locally, on behalf of the public, by officials such as EHOs, TSOs and Public Analysts. They ensure similar baseline standards for all consumers. Now, however, the major food retailers voluntarily regulate their own systems at their own expense, promoting individual choice based on their own hierarchy of quality definitions. Food quality becomes not so much a binary definition under these regulatory conditions but rather one which absorbs constant redefinitions evolved by the peculiar powers society and government have bestowed on retailers as the new custodians of the food system as a whole.

A key element in the ability of the major retailers to engage in self-regulation is that they work in tandem with public regulatory activities. Three innovations in the 1990s have helped smooth the process between public and private styles of regulation. One of these is the home authority principle. Under this provision, nationwide food retailers are entitled to elect one local authority (usually the one in which the company's head office is located) through whom liaison with local government regulators will take place.

The other, and subsequent, new procedures and techniques of assessment were industry codes of practice and hazard analysis, and both provide further legitimisation for retailers' supply chain management. The former allow the food industry to develop its own guides of compliance with food regulation. The latter has two components: that the frequency of site inspections by EHOs is decided by the food safety 'risk' posed by the business; and that the business itself must show that it has identified the safety hazards in its own operation and has taken steps to control them. The promotion of both codes of practice and of hazard analysis allows EHOs to differentiate, in their regulatory activity, between the 'superleague' and other food retailers. Crucially, it represents an acknowledgement by government, and a willingness of the 'superleague' of retailers themselves, to in large part engage in self-regulation. In other words, it is a key step in retailer-led moves towards private-interest regulation.

In practice, EHOs now increasingly adopt a bifurcated approach to the regulation of food retailing. They largely retain their traditional approach, mixed with some rudimentary hazard analysis, to the bulk of independent traders. For the major retailers, though, EHOs are adopting an auditing approach in which their management systems are tested.

There is moreover a certain naivety of implementation practices amongst most of those who advocate regulatory reform (James 1997; MAFF 1998; House of Commons Agricultural Committee 1998: para. 113), in that they advocate a ratcheting up of enforcement standards and assume that such a wish will become practice. In essence, what these champions of reform wish to do is to reduce the discretion (or what the implementation literature calls 'coping strategies') of regulators. However, there is little or no attempt to try and understand the context in which they operate. Yet, Lipsky (1988) has shown that 'street-level bureaucrats' are key actors in the policy process. The behaviour of regulators is closely related to their perceptions and attitudes (Winter 1990: 32). As we show in Chapter 10, there is considerable richness and complexity in the understanding of the regulatory geography of local food regulation. Different priorities and interests come to the fore and so at the local level individual EHOs and their departments will seek to retain discretion – to assist in their coping with those they regulate – to suit their own values and serve their own purposes. There are limits to public regulation just as much as there are to the competitive opportunism of private interests in food supply.

One reason why it is important for local regulators to retain discretion is to cope with the diverse range of food retailers that they regulate and thus the different situations in which they find themselves. For example, EHOs are likely to be a source of knowledge and authority on food law and its implementation to small, independent retail outlets, and a relationship between the two will develop accordingly. However, amongst some of the EHOs with whom we spent time, as well as some of the local store managers who were subsequently interviewed, there is a clear sense that the balance of knowledge relating, for example, to the introduction of new rules or of management systems is tilting ever further towards the corporate retailers and away from officials. During a visit with a senior EHO to a major corporate retailer in Newton, the store manager made it clear that he was able to 'educate' the EHO about food hygiene standards. A common perception amongst EHOs is that the larger firms have the resources to ensure that their staff are kept well informed. As one local official in Newton explained, when talking of the corporate retailers, 'they do employ their own EHOs and when new legislation comes up, they make sure, it's their job, that they are really up there with the latest information, and they can advise their firm accordingly.'

For the Food Standards Agency, therefore, it would be a mistake to simply call for greater uniformity in the regulatory practices of EHOs or more generally assume that broad 'political will' can progress a more publicly accountable food system. The current constitution of the retailing sector and the diversity in its management of food quality means that EHOs too are having to act according to the circumstances of different retail outlets. The greater challenge is to link together the regulatory practices of EHOs in relation to independents and others in the 'lower tier' of retailing (which may be well suited to a geographically-based system of regulation via local authorities) with the supply chain and management system regulation of the major supermarkets. For the latter, who regulate their own supply chains at their own cost, geographical forms of regulation appear increasingly anachronistic. Models of dual regulation (i.e. spatial and supply-based) go beyond anything envisaged in the White Paper (MAFF 1998) on the functions of the Food Standards Agency.

Food governance

Such depictions of the prospects bring together the need to appreciate the dynamics of both public and private sector regulation. In doing so, we begin to challenge the conventional analysis of British government which places strong, centralised departments to the fore. As Smith (1998) points out, however, the Westminster model of government faces theoretical and empirical challenges. In particular, traditional models of government fail to adequately explain a shift 'from a more directive state to a more fragmented state [in which there is a more] flexible form of control rather than government as direct control' (Smith 1998: 51). Smith reviews conventional explanations for the rise of governance, which fall into two broad categories. One relates to the emergence and legitimacy of the new right and 'public choice', and the other to internationalisation and globalisation (Smith 1998: 51–7). Through our detailed analysis of the contemporary development of food policy and regulation, we can begin to throw some light on the validity of such a generalised governance model.

The James Report on the Food Standards Agency (1997) painted a picture of (and indeed officially admitted for the first time) fragmented food regulation. A number of public and semi-private bodies performed a range of roles maintaining food standards. It is precisely this sense of government in which knowledge and authority are fragmented which has been so important to the notion of governance (Rhodes 1997). Nevertheless, an analysis of food governance necessarily begins with the role of the Ministry of Agriculture, Fisheries and Food (MAFF). MAFF, because of its responsibilities towards agriculture, food processors, distributors and the consumer, is at the centre of the regulatory network, (though, of course, this situation will change with the establishment of the Food Standards Agency). MAFF's authority, however, has frayed considerably over the years. The Department of Health (DoH) is an increasingly authoritative source of policy and knowledge on food issues. The DoH takes main responsibility for food nutrition and safety issues and so has a more public health and food policy culture than MAFF, and is widely perceived to have a more sympathetic outlook to 'the consumer'. Indeed, our interviews with DoH officials confirmed that they see themselves as much more attuned to consumer needs. The inevitable turf battles between the two departments over food are but one manifestation of MAFF's growing inability to control the environment in which it operates.

It is also important to take a longer term view of MAFF's role and position within the food system. In the early postwar years, government support for agricultural policy was agreed by all political parties and the major agricultural interest groups. With a fair degree of unanimity over the aims and means of agricultural policy, MAFF was able to operate in a stable policy environment. Not only did farmers stand to benefit from existing arrangements, but so too did other elements in the food system. Agricultural suppliers were able to cater for an increasingly intensive agriculture, and distributors, retailers and consumers were able to purchase food of apparently good (if baseline) quality. What food could not be supplied by British farmers was freely imported. Food policy was depoliticised and considered an adjunct of agricultural policy.

As the nature and balance of power within the food system altered and combined with a new political situation so tensions became apparent. For example, as the Conservative government of the 1950s increasingly focused its attention on ensuring efficient production rather than simply maximum output to limit the extent of Exchequer liabilities, so farmers began to find their incomes squeezed. Farmers and the Labour Party began to complain at what they regarded as the unproductive distributors who, it was claimed, were making excessive profits. Later, in the early 1970s, entry into the EEC gradually inflamed the situation still further as the excesses of the Common Agricultural Policy helped to politicise domestic agricultural policy. Throughout this period, however, MAFF retained the appearance of controlling its operating environment and farmers were secure in their relationship with the Ministry.

When the crisis in the food system became fully apparent in the 1980s (see Chapters 2 and 4), MAFF proved to be unable to respond to the external demands, either organisationally or by taking executive action. In organisational terms, MAFF can be characterised as a passive department which, like many organisations, is well able to respond to demands from a 'settled environment'. It has very little experience of operating in a situation which it has not chosen to actively control, nor does it seemingly manage to learn such contemporary rudiments of regulatory survival. An important part of its strategy in the developing 'food crisis' has, therefore, been to try to recreate its traditional pattern of relationships at the same time as reluctantly absorbing the necessary realities of retail power. Wherever possible, and for as long as possible, contacts have been limited to trusted groups and negotiations to the usual Whitehall rules of secrecy. 'Outsiders', such as consumer groups, are kept at a distance from core policy debates.

However, the nature of the debate surrounding food quality and hygiene has been such that MAFF's attitude has looked increasingly anachronistic, and it has been portrayed as the 'friend' of the producer and 'enemy' of the consumer. This may not be quite fair, but it has been a social reality nonetheless. As the debate on food standards has moved beyond MAFF's policy environment under its control, to include other departments, consumer groups, local government and other elements in the food system, so its ability to speak authoritatively has been weakened. Not surprisingly, it has frequently found itself on the defensive.

MAFF's weak position with regard to food quality and hygiene has been compounded by its executive limitations. Although ultimately it is the minister who remains the person responsible for food standards, as we have seen, these have to be enforced by local government officials. Moreover, the growth of private-interest regulation, in which retailers regulate their suppliers to their own standards, brings another set of interests into the regulatory apparatus with which government must negotiate, cooperate or seek to coerce. The separation of policy and executive work amongst the different tiers of British government is a standard administrative practice, and one which the Food Standards Agency is proposing to overcome. Nevertheless, it does mean that under the administrative arrangements that prevailed in the last two decades the 'problem' of food hygiene could be interpreted in different ways by different interests. For example, within MAFF it is likely that there was at least a partial attempt to structure the issue and responses in such a way as to defend the minister and chosen interests.

Food retailers, as we have seen in relation to BSE above, were concerned to protect and expand their markets and emphasise the quality and safety of their products by distancing themselves from those upstream. Within local government, as we saw in Chapter 10, the high profile of food hygiene could be used to help safeguard the funding for an existing service. Thus, as Smith (1998: 51) suggests, more generally we should think of new forms of 'governance' as a flexible form of control rather than government as direct control. It is also a form of control that is much more variable than we might expect. Hence, as we see, public-interest and private-interest regulation imply quite different capabilities for both government and consumers.

Conclusion

'Governance', in the general sense, usefully highlights the need to look beyond government departments to understand policy making and its implementation. Food policy is not just a problem for government. While a range of bodies may be involved in the policy process, our work in the food sphere cautions against overemphasising this aspect. Governance does not necessarily 'broaden' the range of interests or substantially change the focus of interests served by the policy process. Rather, it is the specific, pragmatic, reluctant ways in which it does this that become significant. As Grant (1993) reminds us, since the mid-1970s:

> Britain displays many of the characteristics of a company state. In a company state the most important form of business–state contact is the direct one between company and government. Government prioritises such forms of contact over associative intermediation.
>
> (1993: 14)

It is not only at the level of policy formation that such relationships will be found, but also at the level of policy implementation. Our analysis of food regulation certainly lends support to the claim of Williamson (1989) that public and private interests work together in sophisticated networks to secure the implementation of policy goals. When the 'interdependence of government and business is a pervasive reality' (Grant 1993: 31) public and private policy goals and their implementation inevitably become blurred. Groups marginal to the policy process, such as consumer organisations, remain so.

The balance between public and private interests and the marginalisation of consumer groups in the food sector is well illustrated by the debates surrounding the introduction of genetically modified crops into the food chain. Quite clearly, there are major producer interests keen to promote the use of genetically modified foods and to ensure that regulations do not unduly constrain their interests. Consumer groups meanwhile have argued for more stringent testing of genetically modified foods and stricter labelling of products that contain them. The response of the EU in its direc-tive covering the labelling of genetically modified foods has been described in a leader in the *Guardian* as a 'curate's egg'. The leader continued:

> As always in this political climate, it seems that corporate interests have triumphed and the consumer has been fobbed off. The new labels will appear in minute type on very few products and most people will assume that food not labelled as containing GM foods will be GM-free. In fact, GM soya and maize, and GM lecithins, or thickeners, may already be used in more than 60 per cent of all processed or packaged foods, but these may escape the new labelling laws because the manufac-turing process can render them unidentifiable.
>
> (*Guardian*, 1 September 1998)

Where public-interest regulation is seen to be wanting, then private-interest regulation may step in, where it senses a potential market opportunity. The Iceland frozen foods chain, with less than 2 per cent of the food market and under considerable pressure from its larger retailer rivals, has undertaken a number of initiatives to try and define its place in the market. One is to guarantee that no Iceland own-brand product manufactured after 1 May 1998 will contain any genetically modified ingredients. The company's market research showed that 81 per cent of its customers were concerned that they could be buying GM food which they would otherwise avoid. The company is only able to do this because it can be sure that 'our suppliers are already using guaranteed non-modified' products. Iceland has, therefore, potentially stolen a march on Tesco, Safeway and Sainsbury's, who will obvi-ously be closely monitoring the situation to see if they need to also use their supply chains to source non-GM products. So, while governments have not regulated to make non-GM food a choice for consumers, retailers have. The case of GM food is highly revealing of the fact that both public-interest and private-interest regulation are about creating and sustaining markets.

Consumer groups, however, have once again been sidelined. Indeed, the opposition to GM foods has not been led by food consumer groups, but by environmental groups which have promoted the consumer perspective. Companies producing GM foods have also focused their attention on the likes of Friends of the Earth and Greenpeace rather than mainstream consumer groups. Environmental groups have shown an ability to mobilise at the grassroots level to protest against the production of GM crops, which is simply beyond the resource capabilities of food consumer groups. The spilling over of environmentalism into other policy arenas can prove to be unsettling for established policy communities. In the case of GM food this may be no more than a temporary incursion, but environmental groups have begun to change the terms of the debate and the means by which it is conducted, to ways which are somewhat different from those of consumer groups. Comparative analysis between these representative groups is overdue. When, as the case of GM foods shows, private-interest regulation has gone further than public-interest regulation in assuaging consumer concerns, the major retailers are unlikely to welcome the activities of environmentalists if they cast doubt on the wider safety of food. Maintaining confidence and trust in what we consume is an integral part of the retailers' strategy, but, as a succession of food scares have shown, constructing and maintaining relationships of trust between the buyers and sellers of food is a long-term and active process.

This more recent example of food controversy indicates how fluid, flexible and, indeed, contingent the character of British retailer-led food governance seems to be – for the time being, at least – in absorbing both the limits of state and the anxieties of consumers. Despite these pressures, the model holds the ability to continuously construct the consumer interest in ways which maintain a maturing compromise between private interests and the state.

Appendix

The social research methods employed in the study

We outline here the variety of methods employed in the study, which took place over a three-year period (1994–7). The material generated from these methods is both referred to directly in the book and has more generally formed the basis of the arguments in it. In Part I of the book, the conceptual discussion has been developed out of the detailed analysis of the evidence, even though little of the evidence is actually cited. Parts II–III refer to selected extracts of the primary and secondary evidence, and it is this that provides the basis of the analysis. It is also important to mention the type of analysis that has been conducted on this evidence. The ethnographic material has been fully read by the researchers, and has been subject to qualitative keyword analysis on the basis of the key parameters of the study. The interviews at the local and strategic level were fully assessed and then key quotations selected which represented the full responses received. This approach was also used with the field diaries and field notes. In the writing up of the material, it has been necessary to match the secondary and primary evidence. Reference is made in the text to the sourcing of this material.

The research methods followed four interrelated lines. These concern:

- the development of a documentary data base on British food organisations, including retailers and regulatory organisations;
- strategic policy/private sector interviewing of key agencies;
- participant observation techniques on professional groups engaged in the implementation of policy;
- policy discussions and follow up interviews in the discussion of results at the national and European level.

Strategic level interviews

Over fifty interviews were conducted at the strategic level with subjects including leading executives and officials in food retail businesses, national government food regulatory bodies and consumer groups. Access was successfully achieved through detailed preparation, through the use of documentary sources and in careful planning of the interview schedules. In many cases, the agreement to interview and the location of the most effective respondent, once agreement had been reached, could take a period of months to arrange. For the retailers, board approval was usually sought before interviews were agreed. Overall, only a minority of organisations refused to be involved. The development of our documentary data base and the successful completion of interviews and transcriptions represents a 'live' research resource.

The primary evidence gained from the strategic-level interviews (involving corporate retailers, government officials in the Ministry of Agriculture, Fisheries and Food (MAFF), Department of Health and the Department of Trade and Industry, trade associations and consumer and professional bodies) has focused on four areas related to the objectives of the research. These are:

1 the different and often competing perspectives, definitions and approaches to food quality and the provision of food choices;
2 how the different organisations – and particularly some of the key actors inside them – interact in their activities and priorities; what strategies they adopt; how they view their role in providing food choices and how they attempt to construct the consumer interest;
3 the degree to which they are involved in contemporary food policy making and development;
4 the extent to which the goals set out in key policy statements, like the *Health of the Nation* White Paper, and the Food Safety Act 1990, are being realised by the different organisations and professional groups.

The large majority of interviews were tape recorded and all have been transcribed. This provides a detailed, insightful and unique body of evidence at the strategic level of national food policy making and regulation.

National and local implementation: participant observation methods

Primary and secondary evidence was collected from two contrasting local authorities. One was an inner-London borough, the other a county in the west of England. For each of the fieldwork sites, the same multimethod procedures were followed in order to collect data. These included:

1 the description of the context for food retail provision in the locality, both from observation and from secondary material obtained from the local authority;
2 the review of the historical development of these retail patterns (evidence gained from the local planning offices and interviews with key personnel); future developments based on retail development proposals and planning policy;
3 a study of the relationship between retailers and local planning officials

based on previous and current planning applications as case studies and interviews with key planning personnel;

4 the understanding of the work of food regulation enforcement officers through interviews with key personnel in their coordinating bodies (e.g. LACOTS, Department of Health);

5 the study of the way in which food legislation is enforced through the direct observation of the regulators at work, following EHOs and TSOs as they go about their daily duties, until a point of saturation of information is reached;

6 the study of the micro-relationships between retailers and the regulators through direct observation and formal and informal interviews with shop personnel.

The multi-methodological approach to the research in an inner London borough and a county in the west of England has generated several sources of data that both provide new insights into the construction of the consumer interest at the local level, and supplement the retailing and regulation profiles constructed at the national level. They include:

1 Recorded interviews with key personnel in positions of regulatory responsibility: at the Department of Health; a senior member of staff at LACOTS; the Principal EHOs; the Principal TSOs; Senior Planning Officers responsible for retail planning; and the Public Analyst responsible for the study areas Trading Standards and Environmental Health Departments' investigations. (The Chief EHOs and TSOs were interviewed at the outset and close of the fieldwork period.) These interviews describe the manner in which food retail is regulated in the study areas, and reflect upon the national and, where appropriate, European situation. They also consider the nature of the relationship between local, national and international food regulatory systems. Therefore, in addition to providing a perspective on the local articulation of the consumer interest, they also complement the national work with an illustration of the tensions between the three tiers of public policy.

2 The direct and participant observation of the work of EHOs and TSOs in the study areas, undertaken both inside the council offices and throughout the borough/district, and varied according to the day-to-day responsibilities of the officers involved. Work in the office environments involved the observation of activities and responsibilities and a study of the monitoring systems employed by local authorities. Importantly, it included detailed conversations about the nature of the operation of the departments as a whole, and the opinions of personnel on many issues relating to food quality control. The nature of the work outside the office was dictated by day-to-day schedules of the officers; however, some bias was created by their attempts to ensure that a wide range of food retail establishments were visited. These included large supermarkets, small independent shops,

restaurants and cafes, food wholesalers and market traders. There were, there-fore, three types of data generated by this ethnographic work: studies of the nature of local government food regulation and its varying application to different forms of retail establishments, a narrative on the nature of food retailing in the study area, and case study material illustrative of the different types of food retail as reflective of national trends.

Most of the research comprised the direct observation of professional behaviour within the council departmental offices and, critically, in food premises themselves during inspection. This observation became increas-ingly participative as the field research progressed, allowing a greater understanding of inspection procedures not just through 'on the job' learning but also from a basic environmental health food safety training course and examination. It was left to the officer undertaking the site visit to decide how best they felt they should explain the presence of their companion: sometimes the real identity and full purpose was disclosed, but usually a loose guise of 'colleague-in-training' was adopted. No objec-tions were ever made on the part of the food retailer to the presence of the researcher. On those occasions when the retailer was aware of the purpose of the research, much discussion would usually be entailed regarding their opinions of the regulatory process. Local (both corporate and inde-pendent) retailers were also approached and interviewed separately.

The use of an audio-tape recorder was judged to be inappropriate for this part of the research, because of the constant interaction with members of the public who were not aware of the researcher's identity. Instead, a field diary was kept throughout and the details of various visits to food premises, as well as information gleaned from time spent within the local authority offices, was recorded in this way. Secondary material was also collected. This included both educative material produced by national regulatory bodies for local EHOs and TSOs, and educative mate-rial produced by the local authority departments for traders.

Despite the depth and range of research material gained, it was clear that officers were instructed not to take the researcher to places where they could expect an unfriendly reception or where standards were very poor and might reflect negatively the effectiveness of their work, or as one TSO office explained, place the researcher in a situation where she would be required to give court evidence if a prosecution was necessary. There were also food premises that women officers did not visit because of the perceived risk of violence. But despite this, the field diaries upon which this chapter is based (one for each study site) review more than sixty site visits *in situ*, and many more anecdotes from the officers themselves regarding past experiences.

3   Detailed interviews conducted with the managers of large retailing estab-lishments. These investigated the relationship between the store and local food law enforcement officers, the system of quality control used by the store, and the relationship between the store and its own head office.

Information was also gathered on local food retail change, and company developments both within the study areas and nationally. These interviews therefore not only contributed to the study of local food retail and regulation, but provided case study material illustrative of the local articulation of the national retailing themes deciphered from the strategic and national level primary research. A range of retail stores agreed to be interviewed, including Kwik Save, Marks and Spencer, Sainsbury, Safeway, Gateway, Iceland and Tesco.

4 A study of retail planning. In addition to the interviews noted earlier, an examination of recent retail planning materials was undertaken in the planning office in the study areas.

5 The collection of secondary materials. A range of material was collected that included local authority planning documents, socioeconomic documentation and, as noted, materials received and generated by the Environmental Health and Trading Standards Departments.

Bibliography

Allaire, B. and Boyer, R. (1995) *La grande transformation de l'agriculture: lectures conventionannalistes et regulationnistes*, Paris: INRA Economica.

Appadurai, A. (1986) 'Introduction: commodities and the politics of value', in *The Social Life of Things: Commodities in Cultural Perspective*, Cambridge: Cambridge University Press, 1–50.

Audit Commission (1990) *The Study of Environmental Health Officer Enforcement*, London: HMSO.

Barnell, H.R., Coomes, T.J. and Hollingsworth, D.F. (1968) 'Some aspects of the implementation of food policy', *Proceedings of the Nutrition Society* 27: 8–13.

Bauman, Z. (1988) *Freedom*, Milton Keynes: Open University Press.

—— (1996) 'From pilgrim to tourist – or a short history of identity', in S. Hall and P. du Gay (eds), *Questions of Cultural Identity*, London: Sage, 18–36.

BEUC (Bureau Européen des Unions de Consommateurs) (1996) *Memorandum to the Dutch Presidency*, BEUC/429/96, Brussels: BEUC.

Booker, C. and North, R. (1994) *The Mad Officials*, London: Constable.

BNF (British Nutrition Foundation) (1995) *Annual Report and Accounts 1994–95*, London: BNF.

BRC (British Retail Consortium) (n.d.) *Annual Report 1993/94: Food and Drink Supplement*, London: BRC.

Bromley, D.W. (1991) *Environment and Economy: Property Rights and Public Policy*, Oxford: Blackwell.

Cabinet Office (1995) *Progress Through Partnership: Retail and Distribution*, Technology Foresight 15, London: HMSO.

Christopherson, S. (1993) 'Market rules and territorial outcomes: the case of the United States', *International Journal of Urban and Regional Research* 17: 274–88.

Clark, G. (1992) ' "Real" regulation: the administrative state', *Environment and Planning A* 24: 615–27.

Commission of the European Communities (1985) *Completion of the Internal Market: Community Legislation on Foodstuffs. (Communication from the Commission to the Council and to the European Parliament)*, COM/85/0603/FIN, Luxembourg: Office for Official Publications of the European Communities.

—— (1997) *The General Principles of Food Law in the European Union: Commission Green Paper*, COM/97/0176/FIN, Luxembourg: Office for Official Publications of the European Communities.

Consumer Congress and National Consumer Council (1994) *Consumer Representation in the Public Sector: How the Ministry of Agriculture, Fisheries and Food and the Benefits*

Agency Use Consumer Representatives, London: Consumer Congress and National Consumer Council.

Consumers in the European Community Group (1992) *Briefing Paper on the Proposed Reform of the Scientific Committee for Food*, London: Consumers in the European Community Group.

Cox, G., Lowe, P. and Winter, M. (1986) 'From state direction to self-regulation: the historical development of corporatism in British agriculture', *Policy and Politics* 14: 475–90.

Deleuze, G. (1992) 'Postscript on the societies of control', *October* 59: 3–7.

Department of Health (1992) *The Health of the Nation: A Strategy for Health in England*, Cm. 1986, London: HMSO.

—— (1994a) *Eat Well! An Action Plan from the Nutrition Task Force to Achieve the Health of the Nation Targets on Diet and Nutrition*, London: Department of Health.

—— (1994b) *Nutritional Aspects of Cardiovascular Disease: Report of the Cardiovascular Review Group Committee on Medical Aspects of Food Policy*, Report on Health and Social Subjects 46, London: HMSO.

Department of the Environment (1993) *Town Centre and Retail Development*, Planning Policy Guidance Note 6, London: HMSO.

de Vroom, Bert (1985) 'Quality regulation in the Dutch pharmaceutical industry: conditions for private regulation by business interest association', in W. Streeck and P.C. Schmitter (eds), *Private Interest Government: Beyond Market and State*, London: Sage, 128–49.

Dobson, B., Beardsworth, A., Keil, T. and Walker, R. (1994) *Diet, Choice and Poverty: Social, Cultural and Nutritional Aspects of Food Consumption among Low-Income Families*, London: Family Policy Studies Centre.

Doel, C. (1996) 'Market development and organisational change: the case of the food industry', in N. Wrigley and M.S. Lowe (eds), *Retailing, Consumption and Capital: Toward a New Retail Geography*, Harlow: Longman, 48–68.

Euro PA and Associates (1998) *The UK Food amd Drink Industry*, Cambridge: Northborough.

European Commission (1997) 'Green Paper on Commerce', *Bulletin of the European Union*, 2/97.

Fine, B. (1993) 'Modernity, urbanism and modern consumption – a comment', *Environment and Planning D. Society and Space* 11: 599–601.

Fine, B., Heasman, M. and Wright, J. (1996) *Consumption in the Age of Affluence: The World of Food*, London: Routledge.

Fine, B. and Leopold, E. (1993) *The World of Consumption*, London: Routledge.

Flynn, A., Harrison, M. and Marsden, T. (1998) 'Regulation rights and the structuring of food choices', in A. Murcott (ed.), *'The Nation's Diet': The Social Science of Food Choice*, Harlow: Addison Wesley Longman, 152–67.

Flynn, A., Lowe, P. and Winter, M. (1996) 'The political power of farmers: an English perspective', *Rural History* 7: 15–32.

Flynn, A. and Marsden, T. (1992) 'Food regulation in a period of agricultural retreat: the British experience', *Geoforum* 23: 85–93.

—— (1995) 'Rural change, regulation and sustainability', *Environment and Planning A* 27: 1179–80.

Flynn, A., Marsden, T. and Ward, N. (1991) 'Managing food? A critical perspective on the British experience', *INRA Acte et Communications* 7: 159–81.

—— (1994) 'Retailing, the food system and the regulatory state', in P. Lowe, T.K. Marsden and S. Whatmore (eds), *Regulating Agriculture*, London: Wiley.

Food and Drink Federation (n.d.) *Annual Report '95*, London: Food and Drink Federation.

Foreman, S. (1989) *Loaves and Fishes*, London: HMSO.

Gabriel, Y. and Lang, T. (1995) *The Unmanageable Consumer: Contemporary Consumption and its Fragmentation*, London: Sage.

Glennie, P. and Thrift, N. (1993) 'Modern consumption: theorising commodities and consumers', *Environment and Planning D. Society and Space* 10: 423–43.

—— (1996) 'Consumption, shopping and gender', in N. Wrigley and M.S. Lowe (eds), *Retailing, Consumption and Capital: Toward a New Retail Geography*, Harlow: Longman.

Goodman, D. and Redclift, M. (1991) *Refashioning Nature: Food, Ecology and Culture*, London: Routledge.

Goodman, D. and Watts, M. (1994) 'Reconfiguring the rural or fording the divide? Capitalist restructuring and the global agro-food system', *Journal of Peasant Studies* 22: 1–49.

Granovetter, M. (1992) 'Economic institutions as social constructions. A framework for analysis', *Acta Sociologica* 35: 3–11.

Grant, W. (1983) 'The National Farmers' Union: the classic case of incorporation', in D. Marsh (ed.), *Pressure Politics*, London: Junction Books, 129–43.

—— (1987) 'Introduction', in W. Grant (ed.), *Business Interests, Organizational Development and Private Interest Government*, Berlin: Walter de Gruyter, 1–17.

—— (1993) *Business and Politics in Britain*, 3rd edn, Basingstoke: Macmillan.

Gray, P.S. (1990) 'Food law and the internal market', *Food Policy* 15: 118–21.

Gray, P.S. (1993) '1993 and European Food Law, an end or a new beginning?', *European Food Law Review* 1–2: 1–16.

Guy, C. (1995) *The Retail Development Process: Location, Property and Planning*, London: Routledge.

—— (1998) 'Controlling new retail spaces: the impress of planning policies in Western Europe', *Urban Studies*, vol. 35, nos 5–6: 953–79.

Hancher, L. and Moran, M. (1989) 'Organising regulatory space', in L. Hancher and M. Moran (eds), *Capitalism, Culture and Economic Regulation*, Oxford: Clarendon Press.

Harrison, M., Flynn, A. and Marsden, T. (1997) 'Contested regulatory practice and the implementation of food policy: exploring the local and national interface', *Transactions of the Institute of British Geographers* 22: 473–87.

Harrison, M.L. (1991) 'Citizenship, consumption and rights: a comment on B.S. Turner's theory of citizenship', *Sociology* 25: 209–13.

House of Commons Agriculture Committee (1998) *Fourth Report: Food Safety*, HC 331, London: The Stationery Office.

Hughes, A. (1996) 'Transforming power relationships between food retailers and brand manufacturers in the UK and USA', paper presented at the 3rd CIRASS/EIRASS International Conference on Retailing and Services Science, Telfs/Buchen, Austria.

Hutter, B.M. (1986) 'An inspector calls: the importance of proactive enforcement in the regulatory context', *British Journal of Criminology* 26: 114–28.

—— (1988) *The Reasonable Arm of the Law? The Law Enforcement Procedures of Environmental Health Officers*, Oxford: Clarendon Press.

—— (1989) 'Variations in regulatory enforcement styles', *Law and Policy* 11: 154–74.

Hyde, S., Balloch, S. and Ainley, P. (1989) *A Social Atlas of Poverty in Lewisham*, London: Centre for Inner City Studies, Goldsmiths College.

Institute of Grocery Distribution (1996) 'Consumer attitudes to genetically modified foods', paper presented at British Nutrition Foundation Conference Biotechnology: Science and the Consumer.

Institute of Public Policy Research (1994) *Off Our Trolleys: Food Retailing and the Hypermarket Economy*, London: Institute of Public Policy Research.

James, P. (1997) 'Food Standards Agency: an interim proposal', unpublished report, Rowett Research Institute, Aberdeen.

Jones, P. (1994) *Rights*, Basingstoke: Macmillan.

Jukes, D. (1992) 'Food law and regulation: is the consumer voice heard?', in National Consumer Council (ed.), *Your Food: Whose Choice?*, London: HMSO, 157–77.

Keane, A. and Willetts, A. (1995) *Concepts of Healthy Eating: An Anthropological Investigation in South East London*, London: Goldsmiths College.

Lacey, R. (1991) *Unfit for Human Consumption: Food in Crisis. The Consequences of Putting Profit Before Safety*, London: Souvenir Press.

LACOTS (Local Authorities Coordinating Body on Food and Trading Standards) (1993) 'Food safety enforcement and prosecution policies. A LACOTS secondee project on local authority food safety enforcement and prosecution policies for distribution to Chief Environmental Health Officers', restricted access document, LACOTS, Croydon.

—— (1994a) *Guidance on Food Safety and Enforcing Policies: Guidance for Local Authority Environmental Health Departments*, Croydon: LACOTS.

—— (1994b) *The Home Authority Principle*, Croydon: LACOTS.

Langston, P., Clarke, G.P. and Clarke, D.B. (1995) *Retail Saturation, Retail Location, and Retail Competition: An Analysis of British Grocery Retailing*, Working paper 95/11, School of Geography, University of Leeds.

—— (1997) 'Retail saturation, retail location, and retail competition: an analysis of British grocery retailing', *Environment and Planning A* 29: 77–10.

—— (1998) 'Retail saturation: the debate in the mid-1990s', *Environment and Planning A* 30: 49–67.

Le Heron, R. and Roche, M. (1995) 'A "fresh" place in food's space', *Area* 27: 23–33.

Lipsky, M. (1988) *Street-Level Bureaucracy: Dilemmas of the Individual in Public Services*, New York: Russell Sage Foundation.

Lury, C. (1996) *Consumer Culture*, Cambridge: Polity Press.

MAFF (Ministry of Agriculture, Fisheries and Food) (1990) *Food Safety Act 1990*, London: HMSO.

—— (1995) *Consumer Panel Annual Report 1994*, CP(95)21/1, London: MAFF.

—— (1998) *The Food Standards Agency: A Force for Change*, Cm. 3830, London: The Stationery Office.

MAFF Food Safety Directorate (1996) *Food Law*, London: MAFF.

Mann, M. (1987) 'Ruling class strategies and citizenship', *Sociology* 21: 339–54.

Marsden, T. and Arce, A. (1995) 'Constructing quality: emerging food networks in the rural transition', *Environment and Planning A* 27: 1261–99.

Marsden, T., Flynn, A. and Harrison, M. (1997) 'Retailing, regulation, and food consumption: the public interest in a privatized world?', *Agribusiness* 13: 211–26.

Marsden, T., Harrison, M. and Flynn, A. (1998) 'Creating competitive space: exploring the social and political maintenance of retail power', *Environment and Planning A* 30: 481–98.

Marsden, T. and Wrigley, N. (1995) 'Regulation, retailing and consumption', *Environment and planning A* 27: 1899–1912.

—— (1996) 'Retailing, the food system and the regulatory state', in N. Wrigley and M.S. Lowe (eds), *Retailing, Consumption and Capital: Towards the New Retail Geography*, Harlow: Longman, 33–47.

Marshall, T.H. (1963) *Sociology at the Crossroads*, London: Heinemann Educational Books.

—— (1965) *Social Policy in the Twentieth Century*, London: Hutchinson.

—— (1981) *The Right to Welfare and Other Essays*, London: Heinemann Educational Books.

McColl, K. (1992) 'EC food regulation: principles for reform', *Consumer Policy Review* 2: 204–12.

Miller, D. (1995) 'Consumption as the vanguard of history: a polemic by way of an introduction', in D. Miller (ed.), *Acknowledging Consumption: A Review of New Studies*, London: Routledge.

Murcott, A. (ed.) (1998) *The Nation's Diet: The Social Science of Food Choice*, London: Longman.

Murdoch, J. and Marsden, T. (1995) 'The spacialisation of politics: local and national actor spaces in environmental conflict', *Transactions of the Institute of British Geographers* 20: 368–80.

National Consumer Council (1988) *Consumers and the Common Agricultural Policy*, London: HMSO.

National Food Alliance (1995) *Working Together for Better Food: The First and the Second Ten Years*, London: National Food Alliance.

OFT (Office of Fair Trading) (1997) *Competition in Retailing*, research report, London: House of Commons.

Office of Science and Technology (1995) *Progress Through Partnership: Food and Drink*, London: HMSO.

Pennington Group (1997) *Report on the Circumstances Leading to the 1996 Outbreak of Infection with E. coli 0157 in Central Scotland, the Implication for Food Safety and the Lessons to be Learned*, Edinburgh: The Stationery Office.

Perissich, R. (1993) 'Food control in the European Community', *Food Control* 4: 58–60.

Pred, A. (1996) 'Interfusions: consumption, identity and the practice of power relations in everyday life', *Environment & Planning A* 28: 11–25.

Pressman, J. and Wildavsky, A. (1973) *Implementation*, Berkeley, CA: University of California Press.

Rabobank Nederland (1994) *The Retail Food Market: Structure, Trends and Strategies*, Utrecht: Rabobank Nederland Agribusiness Research.

Rhodes, R. (1997) *Understanding Governance: Policy Networks, Governance, Reflexivity and Accountability*, London: Open University Press.

Saunders, P. (1986) *Social Theory and the Urban Question*, 3rd edn, London: Hutchinson.

Saunders, P. and Harris, C. (1990) 'Privatisation and the consumer', *Sociology* 24: 57–75.

Self, P. and Storing, H.J. (1962) *The State and the Farmer*, London: Allen & Unwin.

Smith, M.J. (1990) *The Politics of Agricultural Support in Britain*, Aldershot: Dartmouth.

—— (1991) 'From policy community to issue network: salmonella in eggs and the new politics of food', *Public Administration* 69: 235–55.

—— (1998) 'Reconceptualizing the British state: theoretical and empirical challenges to central government', *Public Administration* 76: 45–72.

Streeck, W. and Schmitter, P.C. (eds) (1985) *Private Interest Government: Beyond Market and State*, London: Sage.

Turner, B. (1990) 'Outline of a theory of citizenship', *Sociology* 24: 189–217.

Warde, A. (1990) 'Introduction to the sociology of consumption', *Sociology* 24: 1–4.

—— (1997) *Consumption, Food and Taste: Culinary Antinomies and Commodity Culture*, London: Sage.

Welsh Office (1998) 'Evidence to the Welsh Affairs Committee Inquiry into the problems facing the livestock industry', in *Second Report*, House of Commons Welsh Affairs Committee. HC 447, London: The Stationery Office.

Williamson, P.J. (1989) *Consumerism in Perspective*, London: Sage.

Winter, S. (1990) 'Integrating implementation research', in D.J. Palumbo and D.J. Calista (eds), *Implementation and the Policy Process: Opening up the Black Box*, Westport, CN: Greenwood Press, 19–38.

Wrigley, N. (1991) 'Is the "golden age" of British grocery retailing at a watershed?', *Environment and Planning A* 23: 1537–44.

—— (1992) 'Antitrust regulation and the restructuring of grocery retailing in Britain and the USA', *Environment and Planning A* 24: 727–49.

—— (1994) 'After the store wars? Towards a new era of retail competition?', *Journal of Retail and Consumer Services* 1: 5–20.

—— (1995) 'Retailing and the arbitrage economy: market structure, regulatory frameworks, investment regimes, and spatial outcomes', paper prepared for Harold Innes Centenary Conference, University of Toronto, Canada, 1994; updated 1995.

—— (1998) 'How British retailers have shaped food choice', in A. Murcott (ed.) *The Nation's Diet: The Social Science of Food Change*, London: Longman, 112–29.

Wrigley, N. and Lowe, M.S. (eds) (1996) *Retailing, Consumption and Capital: Toward a New Retail Geography*, Harlow: Longman.

Index